(*...tinued on back*)

X-Ray Fluorescence Spectrometry

CHEMICAL ANALYSIS

A SERIES OF MONOGRAPHS ON
ANALYTICAL CHEMISTRY AND ITS APPLICATIONS

VOLUME 99

WILEY

A WILEY-INTERSCIENCE PUBLICATION

JOHN WILEY & SONS

New York / Chichester / Brisbane / Toronto / Singapore

X-Ray Fluorescence Spectrometry

AEIA

RON JENKINS

International Centre for Diffraction Data
Swarthmore, Pennsylvania

WILEY

A WILEY-INTERSCIENCE PUBLICATION

JOHN WILEY & SONS

New York / Chichester / Brisbane / Toronto / Singapore

Library of Congress Cataloging in Publication Data:

Jenkins, Ron, 1932–
X-ray fluorescence spectrometry.
"A Wiley-Interscience publication."
Includes bibliographies and index.
1. X-ray spectroscopy. I. Title.
QD96.X2J47   1988   543′.08586   88-10797
ISBN 0-471-83675-3

Printed in the United States of America

10 9 8 7 6 5 4 3 2 1

FOREWORD

It is now nearly 30 years since the publication, in 1959, of the Wiley-Interscience monograph *X-Ray Spectrochemical Analysis* by Verne Birks. In the intervening years the X-ray fluorescence method has come through the birth pains of innovation, has survived the early frustrations of application, and has achieved the status of a reliable, fast, accurate and versatile analytical method.

The analytical chemist of today has a vast array of different techniques available for the analysis and characterization of materials, and most would agree that among the more powerful and flexible of these methods are those based on the use of X-ray fluorescence spectrometry. The X-ray fluorescence method is a means of qualitatively and quantitatively determining elements by measurement of the wavelengths and intensities of characteristic emissions. The technique is applicable to all but the very low atomic number elements, with sensitivities down to the low part per million level.

In the late 1950s the elements covered by the X-ray fluorescence method ranged from the higher atomic numbers down to titanium ($Z = 22$). By the mid 1960s the advent of first the ethylene diamine d-tartrate (EDDT) crystal and then the penta-erythritol (PE) crystal, along with the chromium and rhodium anode X-ray tubes, increased the coverable atomic number range to include all elements down to and including aluminum ($Z = 13$). Under certain circumstances even magnesium and sodium were measurable—albeit with rather poor sensitivity. As we entered the mid 1980s the advent of layered synthetic microstructures (LSM's) has allowed measurements down to carbon ($Z = 6$) with fair sensitivity, and even boron at concentration levels of several percent. The sensitivity of the X-ray fluorescence method for the determination of small quantities of material has also improved signifi-

v

cantly. A "small" sample in the late 1950s and early 1960s was typically of the order of milligrams. Today, use of synchrotron or proton source excitation, along with total reflectance geometry, allows measurements at the picogram level.

For some, it is difficult to imagine development at the same exciting level over the next two decades. Many believe that X-ray fluorescence has come as far as it will. I personally do not subscribe to this view. I believe that the problems of rapid and efficient sample homogenization will soon be solved. The development of room temperature solid state detectors has much still to yield. Use of the synchrotron is beginning to reveal areas of application of X-ray spectrometry hitherto not even considered. The use of the personal computer has yet to find its full exploitation in automating both quantitative and qualitative analysis. The development of combination X-ray diffractometer–spectrometers is at last beginning to show fruit. Present indications are that X-ray fluorescence spectrometry will continue to be an exciting and dynamic discipline.

CONTENTS

X-Ray Fluorescence Spectrometry

CHAPTER

1

PRODUCTION AND PROPERTIES OF X-RAYS

1.1 INTRODUCTION

X-rays are a short wavelength form of electromagnetic radiation that was discovered by Wilhelm Roentgen in 1895 [1]. X-ray based techniques provided important tools for the theoretical physicist in the first half of this century, and since the early 1950s they have found an increasing use in the field of materials characterization. Today, methods based on absorptiometry play a vital role in industrial and medical radiography. The simple X-ray field units that were employed at the front in World War I saved tens of thousands of lives [2]. Today the technology has advanced to a high degree of sophistication. Modern X-ray tomographic methods give us an almost complete three-dimensional cross section of the human body, offering an incredibly powerful tool for the medical field. In addition, the analytical techniques based on X-ray diffraction and X-ray spectrometry, both of which were first conceived almost 70 years ago, have become indispensable in the analysis and study of inorganic and organic solids. Today, data obtained from X-ray spectrometers are being used to control steel mills, ore flotation processes, cement kilns, and a whole host of other vital industrial processes (see e.g. ref. 3). X-ray diffractometers are used for the study of ore and mineral deposits, in the production of drugs and pharmaceuticals, and in the study of thin films, stressed and oriented materials, and phase transformations, plus a myriad of other applications in pure and applied research.

X-ray photons are produced following the ejection of an inner orbital electron from an irradiated atom and subsequent transition of atomic orbital electrons from states of high to low energy. When a monochromatic beam of X-ray photons falls onto a specimen, three basic phenomena may result:

1

scatter, absorption or fluorescence. The coherently scattered photons may undergo subsequent interference, leading in turn to the generation of diffraction maxima. The angles at which the diffraction maxima occur can be related to the spacings between planes of atoms in the crystal lattice, and hence X-ray generated diffraction patterns can be used to study the structure of solid materials. Following the discovery of the diffraction of X-rays by Max von Laue in 1913 [4], the use of this method for materials analysis has become very important both in industry and research, to the extent that today it is one of the most useful techniques available for the study of structure dependent properties of materials.

X-ray powder diffractometry involves characterization of materials by use of data that is dependent on the atomic arrangement in the crystal lattice [see e.g. ref. 5]. The technique uses single phase or multiphase (i.e. multicomponent) specimens comprising a random orientation of small crystallites, each of the order of 1–50 μm in diameter. Each crystallite in turn is made up of a regular, ordered array of atoms. An ordered arrangement of atoms (the crystal lattice) contains planes of high atomic density, which in turn means planes of high electron density. A monochromatic beam of X-ray photons will be scattered by these atomic electrons, and if the scattered photons interfere with each other, diffraction maxima may occur. In general, one diffracted line will occur for each unique set of planes in the lattice. A diffraction pattern is typically in the form of a graph of diffraction angle (or interplanar spacing) vs diffracted line intensity. The pattern is thus made up of a series of superimposed diffractograms, one for each unique phase in the specimen. Each of these unique patterns can act as an empirical "fingerprint" for the identification of the various phases, using pattern recognition techniques based on a file of standard single phase patterns. Quantitative phase analysis is also possible, albeit with some difficulty, because of various experimental and other problems, not the least of which is of the large number of diffraction lines occurring from multiphase materials.

A beam of X-rays passing through matter will be attenuated by two processes, scattering and photoelectric absorption. In the majority of cases the greater of these two effects is absorption, and its magnitude, i.e. the fraction of incident X-ray photons lost in passing through the absorber, increases significantly with the average atomic number of the absorbing medium (to a first approximation, as the cube of the atomic number). Thus, when a polychromatic beam of X-rays is passed through a heterogeneous material, areas of high average atomic number will attenuate the beam to a greater extent than areas of lower atomic number. Thus the beam of radiation emerging from the absorber has an intensity distribution across the irradiation area of the specimen, which is related to the average atomic number distribution across the same area. It is upon this principle that all methods of X-ray radiography are based. Study of materials by use of the X-ray absorption process is the oldest of all of the X-ray methods in use, and Roentgen himself included a radiograph of his wife's hand in his first published X-ray paper. Today, there are many different forms of X-ray absorptiometry in use, including industrial radiography, diagnostic medical and dental radiography, and security screening.

Secondary radiation produced from the specimen is characteristic of the elements making up the specimen. The technique used to isolate and measure individual characteristic wavelengths is called X-ray spectrometry. X-ray spectrometry also has its roots back in the early part of this century, stemming from Moseley's work in 1913 [6]. The technique uses either the diffracting power of a single crystal to isolate narrow wavelength bands, or a proportional detector to isolate narrow energy bands, from the polychromatic beam characteristic radiation excited in the sample. The first of these methods is called wavelength dispersive spectrometry, and the second, energy dispersive spectrometry. Because the relationship between emission wavelength and atomic number is known, isolation of individual characteristic lines allows the unique identification of an element to be made, and elemental concentrations can be estimated from characteristic line intensities. Thus this technique is a means of materials characterization in terms of chemical composition.

Although the major thrust of this monograph is to review X-ray spectroscopic techniques, these are by no means the only X-ray based methods that are used for materials analysis and characterization. In addition to the many industrial and medical applications of diagnostic X-ray absorption methods already mentioned, X-rays are also used in areas such as structure determination based on single crystal techniques, space exploration and research, in lithography for the production of microelectronic circuits, and so on.

One of the major limitations with the further development of new X-ray methods is the inability to "focus" X-rays, as can be done with visible light rays. Although it is possible to partially reflect X-rays at low glancing angles, or diffract an X-ray beam with a single crystal, these methods cause significant intensity loss, and fall far short of providing the high intensity, monochromatic beam that would be ideal for, say, an X-ray microscope. Use of synchrotron radiation offers the potential of an intense, highly focused, coherent X-ray beam, but has practical limitations due to size and cost. The X-ray laser could, in principle, provide an attractive alternative, and since the discovery of the laser in 1960, the possibilities of such an X-ray laser have been discussed. Although major research efforts have been, and are still being, made to produce laser action in the far ultraviolet and soft X-ray regions, production of conditions to stimulate laser action in the X-ray region with a net positive gain is difficult, due mainly to the rapid decay rates and high absorption cross sections that are experienced in practice.

1.2 CONTINUOUS RADIATION

When a high energy electron beam is incident upon a specimen, one of the products of the interaction is an emission of a broad wavelength band of radiation called "the continuum." This continuum, which is also referred to as *white radiation* or *bremsstrahlung*, is produced as the impinging high energy electrons are decelerated by the atomic electrons of the elements making up the specimen. A typical intensity–wavelength distribution of this radiation is illustrated in figure 1-1 and is typified by a minimum wavelength λ_{min}, which is roughly proportional to the maximum accelerating potential V of the electrons, i.e. 12.4/V keV. However, at higher potentials, an experimentally measured value of the minimum wavelength of the continuum may yield a somewhat longer wavelength than would be predicted from the effect of Compton scattering. The intensity distribution of the continuum reaches a maximum, I_{max}, at a wavelength 1.5 to 2 times greater than λ_{min}. Increasing the accelerating potential causes the intensity distribution of the continuum to shift towards shorter wavelengths. The curves given in figure 1-1 are for the element molybdenum ($Z = 42$). Note the appearance of characteristic lines of Mo Kα ($\lambda = 0.71$ Å) and Mo Kβ ($\lambda = 0.63$ Å), once the K shell excitation potential of 20 keV has been exceeded. Most commercially available spectrometers utilize a sealed X-ray tube as an excitation source, and these tubes typically employ a heated tungsten filament as a source of electrons and a layer of pure metal (such as chromium, rhodium or tungsten) as the anode. The broad band of white radiation produced by this type of tube is ideal for the excitation of the characteristic lines from a wide range of atomic numbers. In general, the higher the atomic number of the anode material, the more intense the beam of radiation produced by the tube. Conversely, however, because the higher atomic number anode elements generally require thicker exit windows in the tube, the longer wavelength

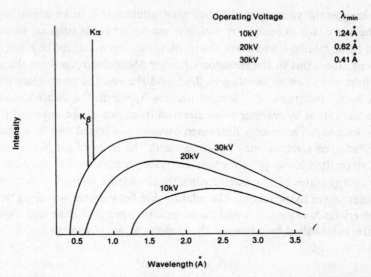

Operating Voltage	λ_{min}
10kV	1.24 Å
20kV	0.62 Å
30kV	0.41 Å

Figure 1-1. Intensity output from a molybdenum target X-ray tube at 10, 20 and 30 kV.

output from such a tube is rather poor, and so these high atomic number anode tubes are less satisfactory for the excitation of longer wavelengths from low atomic number samples (see also section 4.2).

1.3   CHARACTERISTIC RADIATION

If a high energy particle, such as an electron, strikes a bound atomic electron, and the energy E of the particle is greater than the binding energy of the atomic electron, it is possible that the atomic electron will be ejected from its atomic position, departing from the atom with a kinetic energy $(E - \phi)$ equivalent to the difference between that of the initial particle and the binding energy ϕ of the atomic electron. The ejected electron is called a photoelectron and the interaction is referred to as the *photoelectric effect*. While the fate of the ejected photoelectron has little consequence as far as the production and use of characteristic X-radiation from an atom is concerned, it should be mentioned that study of the energy distribution of the emitted photoelectrons gives valuable information about bonding and atomic structure [7]. Study of such information forms the basis of the technique of *photoelectron spectroscopy* (see e.g. ref. 8).

As long as the vacancy in the shell exists, the atom is in an unstable state, and there are two processes by which it can revert to its original state. The first of these involves a rearrangement that does not result in the emission of X-ray photons, but in the emission of other photoelectrons from the atom. The effect is known as *the Auger effect*, and the emitted photoelectrons are called Auger electrons. The second process by which the excited atom can regain stability is by transfer of an electron from one of the outer orbitals to fill the vacancy. The energy difference between the initial and final states of the transferred electron may be given off in the form of an X-ray photon. Since all emitted X-ray photons have energies proportional to the differences in the energy states of atomic electrons, the lines from a given element will be characteristic of that element. The relationship between the wavelength λ of a characteristic X-ray photon and the atomic number Z of the excited element was first established by Moseley [9]. Moseley's law is written

$$1/\lambda = K(Z - \sigma)^2 \qquad (1\text{-}1)$$

in which K is a constant that takes on different values for each spectral series. σ is the shielding constant and has a value just less than unity. The wavelength of the X-ray photon is inversely related to the energy E of the photon according to

$$\lambda = 12.4/E \qquad (1\text{-}2)$$

There are several different combinations of quantum numbers held by the electron in the initial and final states, hence several different X-ray wavelengths will be emitted from a given atom. For those vacancies giving rise to characteristic X-ray photons a series of very simple selection rules can be used to define which electrons can be transferred. For a detailed explanation of these rules refer to section 8.2. Briefly, the principal quantum number n must change by 1, the angular quantum number l must must change by 1 and the vector sum of $l + s$ must be a number changing by $+1$ or 0. In effect, this means that for the K series only p $\rightarrow$ s transitions are allowed, yielding two lines for each principal level change. Vacancies in the L level follow similar rules and give rise to L series lines. There are more of the L lines, since p $\rightarrow$ s, s $\rightarrow$ p and d $\rightarrow$ p transitions are all allowed within the selection rules. In general, electron transitions to the K shell give between two and six K lines, and transitions to the L shell give about twelve strong to moderately strong L lines.

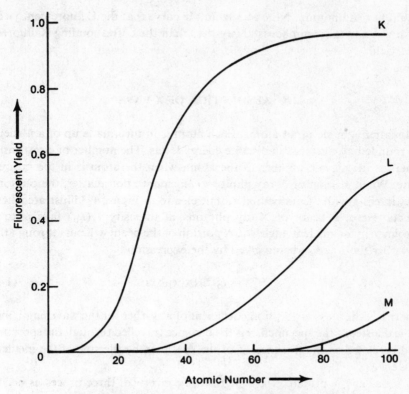

Figure 1-2. Fluorescent yield as a function of atomic number.

Since there are two competing effects by which an atom may return to its initial state, and since only one of these processes will give rise to the production of a characteristic X-ray photon, the intensity of an emitted characteristic X-ray beam will be dependent upon the relative effectiveness of the two processes within a given atom. As an example, the number of quanta of K series radiation emitted per ionized atom is a fixed ratio for a given atomic number, this ratio being called the *fluorescent yield*. Figure 1-2 shows fluorescent yield curves as a function of atomic number. It will be seen that whereas the K fluorescent yield is close to unity for higher atomic numbers, it drops by several orders of magnitude for the very low atomic numbers. In practice this means that if, for example, one were to compare the intensities obtained from pure barium ($Z = 56$) and pure aluminum ($Z = 13$), all other things being equal, pure barium would give about 50 times more counts than would pure aluminum. Note also from the curve that the L fluorescent yield for a given atomic number is always less than the corresponding K fluorescent yield.

1.4   ABSORPTION OF X-RAYS

All matter is made up of atoms. Each atom is in turn made up of a nucleus surrounded by electrons in discrete energy levels. The number of electrons is equal to the atomic number of the atom, when the atom is in the ground state. When a beam of X-ray photons is incident upon matter, the photons may interact with the individual atomic electrons. Figure 1-3 illustrates these effects. Here, a beam of X-ray photons of intensity $I_0(\lambda_0)$ falls onto a specimen at an incident angle ψ_1. A portion of the beam will pass through the absorber the fraction being given by the expression

$$I(\lambda) = I_0(\lambda) \exp(\mu\rho x) \tag{1-3}$$

where μ is the mass absorption coefficient of absorber for the wavelength and ρ the density of the specimen. x is the distance travelled through the specimen and is related to the thickness T of the specimen by the sine of the incident angle ψ_1.

As far as the primary wavelengths are concerned, three processes occur. The first of these is the attenuation of the transmitted beam as described above. Secondly, the primary beam may be scattered over an angle ψ and emerge as coherently and incoherently scattered wavelengths. Thirdly, fluorescence radiation may also arise from the sample. The depth d of specimen contributing to the fluorescence intensity is related to the absorption coefficient of the sample for the fluorescence wavelength and the angle of emergence ψ_2 at which the fluorescence beam is observed.

It will be seen from equation (1-3) that a number $I_0 - I$ of photons have been lost in the absorption process. Although a significant fraction of this loss may be due to scattering, by far the greater loss is due to the photoelectric effect. Photoelectric absorption occurs at each of the energy levels of the atom, thus the total photoelectric absorption is determined by the sum of each individual absorption within a specific shell. Where the absorber is made up of a number of different elements, as is usually the case, the total absorption is made up of the sum of the products of the individual elemental mass absorption coefficients and the weight fractions of the respective

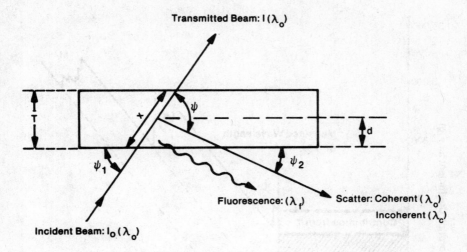

T = Sample Thickness
x = Path Length
d = Penetration Depth
ψ_1 = Incident Angle
ψ_2 = Take-Off Angle
ψ = Scattering Angle
μ = Mass Absorption Coefficient
ρ = Density

- Transmitted Beam Intensity: $I(\lambda_o) = I_o(\lambda_o) \exp(-\mu\rho x)$
- Incoherent Scatter Wavelength: $\lambda_c - \lambda_o = 0.0243[1 - \cos\psi]$
- Fluorescence Penetration Depth: $d = x \sin\psi_2$

Figure 1-3. Interaction of the primary X-ray beam with the sample.

elements. This product is referred to as the total matrix absorption. The value of the mass absorption referred to in equation (1-3) is a function of both the photoelectric absorption and the scattering. However, the influence of photoelectric absorption is usually large in comparison with the scattering and to all intents and purposes the mass absorption coefficient is equivalent to the photoelectric absorption coefficient.

The photoelectric absorption is made up of absorption in the various atomic levels, and it depends on the atomic number. A plot of the mass absorption coefficient as a function of wavelength contains a number of

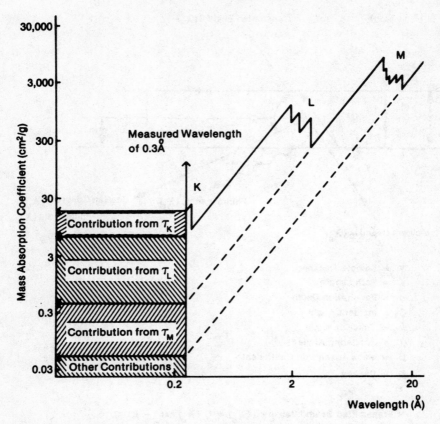

Total Mass Absorption Coefficient for 0.3Å Radiation by Barium = 26 cm²/g

$$\tau_{Total} = \tau_{K_I} + [\tau_{L_I} + \tau_{L_{II}} + \tau_{L_{III}}] + [\tau_{M_I} + \tau_{M_{II}} + \tau_{M_{III}} + \tau_{M_{IV}} + \tau_{M_V}] + others$$

$$= 18 \quad + \qquad 7.5 \qquad + \qquad\qquad 0.45 \qquad\qquad + \quad 0.05$$

$$= 26$$

Figure 1-4. Mass absorption coefficient for barium, showing individual contributions from K, L and M subshells.

discontinuities called *absorption edges*, at wavelengths corresponding to the binding energies of the electrons in the various subshells. As an example, figure 1-4 shows a plot of the mass absorption coefficient as a function of wavelength for the element barium. It will be seen that as the wavelength of the incident X-ray photons become longer, the absorption increases. A number of discontinuities are also seen in the absorption curve: one K discontinuity, three L discontinuities and five M discontinuities. Since it can be assumed that mass absorption coefficient is proportional to the total photoelectric absorption, as indicated at the bottom of the figure, an equation can be written that shows that the total photoelectric absorption is made up of absorption in the K level, plus absorption in the three L levels, and so on. The figure also shows the magnitudes of the various contributions of the levels to the total value of the absorption coefficient of 26 cm^2/g for barium at a wavelength of 0.3 Å. It will be seen the contributions from K, L and M levels respectively are 18, 7.5 and 0.45, with 0.05 coming from other outer levels. If a slightly longer wavelength were chosen, say 0.6 Å, this wavelength would now fall to the long wavelength side of the K absorption edge. This being the case, these photons are insufficiently energetic to eject K electrons from the barium, and since there can be no photoelectric absorption in the level, the term for the K level drops out of the equation. Since this K term is large in comparison with the others, there will be a sudden drop in the absorption curve corresponding exactly to that wavelength at which photoelectric absorption in the K level can no longer occur. Clearly, a similar situation will occur for all other levels, giving the discontinuities in the absorption curve indicated in the figure. Since each atom has a unique set of excitation potentials for the various subshells, each atom will exhibit a characteristic absorption curve. This effect is very important in quantitative X-ray spectrometry, because the intensity of a beam of characteristic photons leaving a specimen is dependent upon the relative absorption effects of the different atoms making up the specimen. This effect is called a matrix effect and is one of the reasons why a curve of characteristic line intensity as a function of element concentration may not be a straight line.

The distances travelled by X-ray photons through solid matter are not very great for the wavelengths and energies of the characteristic lines used in X-ray fluorescence analysis. The x term in equation (1-3) represents the distance travelled, in centimeters, by a monochromatic beam of X-ray photons, in a matrix of average mass absorption coefficient expressed in cm^2/g. The equation can be rearranged in the logarithmic form:

$$2.3 \log_{10} \frac{I}{I_0} = \mu \rho x \tag{1-4}$$

For 99% absorption i.e. $I_0 = 100$ and $I = 1$, equation (1-4) reduces to

$$x(\mu m) = \frac{46,000}{\mu \rho} \tag{1-5}$$

where x is now expressed in microns. Since the densities of most solid materials are in the range of 2 to 7 g/cm^3 and values of mass absorption coefficients are typically in the range 50 to 5,000 cm^2/g, values for x range from a few microns to several hundred microns (see also chapter 7). It will be seen from figure 1-3 that, in the optical arrangement typically employed in commercial X-ray spectrometers, the path length is related to the *depth of penetration d* of a given analyte wavelength:

$$d = x \sin \psi_2 \qquad (1\text{-}6)$$

where ψ_2 is the take-off angle of the spectrometer. Although the depth of penetration is a rather arbitrary measure, it is nevertheless useful, since it does give an indication of the thickness of sample contributing to the measured fluorescence radiation from the specimen for a given analyte wavelength or energy. In most spectrometers the value of ψ_2 is between 30° and 45°, so the value of d is generally about one-half the value of the path length x.

1.5   COHERENT AND INCOHERENT SCATTERING

Scattering occurs when an X-ray photon interacts with the electrons of the target element. Where this interaction is elastic (i.e., no energy is lost in the collision process), the scattering is referred to as coherent (Rayleigh) scattering. Since no energy change is involved, the coherently scattered radiation will retain exactly the same wavelength as that of the incident beam. The origin of the coherently scattered wave is best described by thinking of the primary photon as an electromagnetic wave. When such a wave interacts with an electron, the electron is oscillated by the electric field of the wave and in turn radiates wavelengths of the same frequency as the incident wave. All atoms scatter X-ray photons to a lesser or greater extent, the intensity of the scattering being dependent on the energy of the incident ray and the number of loosely bound outer electrons—in other words, upon the average atomic number. It can also happen that the scattered photon gives up a small part of its energy during the collision, especially where the electron with which the photon collides is only loosely bound. In this instance the scatter is referred to as *incoherent* (Compton scattering). Compton scattering is best presented in terms of the corpuscular nature of the X-ray photon. In this instance, an X-ray photon collides with a loosely bound outer atomic electron. The electron recoils under the impact, removing a small portion of the energy of the primary photon, which is then deflected with the corresponding loss of energy, or increase of wavelength. There is a simple relationship between the incident λ_0 and incoherently scattered wavelength λ_c:

$$\lambda_c - \lambda_0 = 0.0242[1 - \cos \psi] \qquad (1\text{-}7)$$

where ψ is the angle over which the X-ray beam is scattered, which in most commercial spectrometers is equal to 90°. Since the cosine of 90° is zero, there is generally a fixed wavelength difference between the coherently and incoherently scattered lines, equal to around 0.024 Å. This constant difference gives a very practical means of predicting the position of an incoherently scattered line. Also, the coherently scattered line is much broader than a coherently scattered (diffracted) line, because in practice the scattering angle is not a single value, but a range of values, due to the divergence of the primary X-ray beam.

1.6   INTERFERENCE AND DIFFRACTION

X-ray diffraction is a combination of two phenomena—coherent scattering and interference. At any point where two or more waves cross one another, they are said to interfere. Interference does not imply the obstruction of one wavetrain by another, but rather describes the effect of superposition of one wave upon another. The principal of superposition is that the resulting displacement at any point and at any instant may be found by adding the instantaneous displacements that would be produced at the same point by independent wave trains if each were present alone. Under certain geometric conditions, wavelengths that are exactly in phase may add to one another, and those that are exactly out of phase may cancel each other out. Under such conditions, coherently scattered photons may constructively interfere with each other, giving diffraction maxima.

As illustrated in figure 1-5a, a crystal lattice consists of a regular arrangement of atoms, with layers of high atomic density existing throughout the crystal structure. Planes of high atomic density means, in turn, planes of high electron density. Since scattering occurs between impinging X-ray photons and the loosely bound outer orbital atomic electrons, when a monochromatic beam of radiation falls onto the high atomic density layers,

(a)

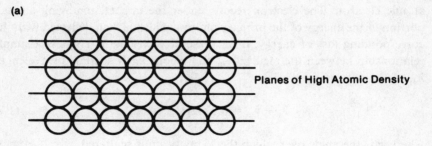

Planes of High Atomic Density

(b)

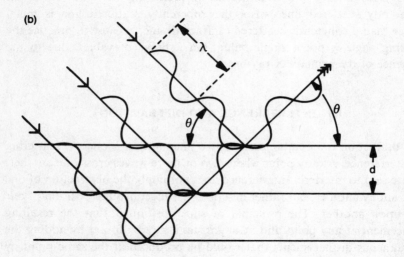

In order to ensure that the scattered waves remain in phase, the path length difference between successive waves (2d.sin θ) must equal a whole number (n) of wavelengths (λ).

i.e. $n\lambda = 2d.\sin\theta$

Figure 1-5.   Diffraction by an ordered array of atoms.

scattering will occur. In order to satisfy the requirement for constructive interference, it is necessary that the scattered waves originating from the individual atoms, i.e. the scattering points, be in phase with one another. The geometric conditions for this condition to occur is illustrated in figure 1-5b. Here, a series of parallel rays strike a set of crystal planes at an angle θ and are scattered as previously described. Reinforcement will occur when the difference in the path lengths of the two interfering waves is equal to a whole number of wavelengths. This path length difference is equal to $ab + bc$, and since $ab = bc = x$, n must equal $2x$ for reinforcement to occur, where n is an integer. It will also be seen that $x = d \sin \theta$, where d is the interplanar spacing; hence the overall condition for reinforcement is that

$$n\lambda = 2d \sin \theta \qquad (1-8)$$

this being a statement of *Bragg's law*. Bragg's law is important in wavelength dispersive spectrometry, since by using a crystal of fixed $2d$, each wavelength will be diffracted at a unique diffraction (Bragg) angle. Thus, by measuring the diffraction angle θ, knowledge of the d-spacing of the analyzing crystal allows the determination of the wavelength. Since there is a simple relationship between wavelength and atomic number, as given by Moseley's law, equation (1-1), one can establish the atomic number(s) of the element(s) from which the wavelengths were emitted.

Figure 1-6 illustrates the origins of peaks and background in a typical experimental X-ray spectrum. The first diagram (a) shows the distribution of the primary exciting radiation, as a plot of intensity vs energy in thousands of volts. This primary spectrum contains both continuum and characteristic lines from the source. As shown in diagram b, scattering effects will cause the primary beam from the source to be scattered by the specimen both coherently and incoherently. The effect of the incoherent scattering is to shift the continuum toward lower energies and to cause broadening of the characteristic lines—again to the lower energy side of the coherently scattered lines. Diagram c shows the positions of the characteristic K lines of an element which is being excited by the primary radiation. In the case of the wavelength dispersive spectrometer (d), the measured spectrum is a composite of diagrams b and c with the addition of the appearance of *harmonics* of each of the characteristic lines, regardless of their origin. The harmonics for lines c and d are indicated by c' and d' for the first order; and c" and d" for the second order. Note that all of the characteristic lines are superimposed on a background. The intensity of the background is mainly dependent upon the

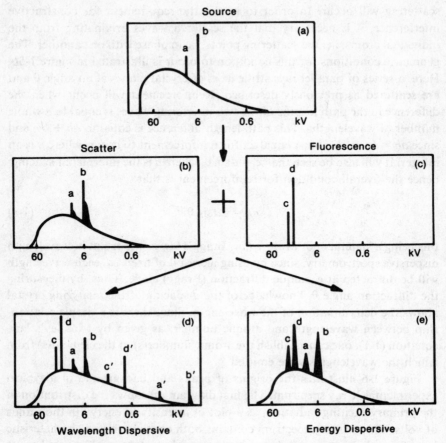

Figure 1-6. Origin of peaks and background in measured X-ray spectra.

scattering power of the sample—the lower the average atomic number, the more efficient the scattering, and therefore the higher the background. As a consequence of this, the peak to background ratio of a given line in a measured spectrum is not just dependent on the concentration of the corresponding element in the specimen, but also on the average atomic number of the specimen. In general, low average atomic number specimens scatter more and absorb less. Thus the background will be high, but so too will be the peak intensity.

REFERENCES

1. Roentgen, W. C. (1898). *Ann. Phys. Chem.* **64**, 1.
2. Nitske, W. Robert, (1971). *The Life of Wilhelm Conrad Röntgen, Discoverer of the X-Ray*, University of Arizona Press.
3. Herglotz, H. K., and Birks, L. S. (1978). *X-Ray Spectrometry*, Dekker: New York.
4. Freidrich, W., Knipping, P., and von Laue, M. (1912). *Ann. Physik* **41**, 971.
5. Klug, H. P., and Alexander, L. E. (1974). *X-Ray Diffraction Procedures*, Wiley: London.
6. Moseley, H. G. J. (1912). *Phil. Mag.* **26**, 1024; (1913) **27**, 703.
7. Bonnelle, C., and Mande C. (1982). "Advances in X-ray spectroscopy." *In* D. J. Nagel (Ed.), *Potential Characteristics and Applications of X-Ray Lasers*, Pergamon: Oxford, England, Chapter 20, pp. 371–410.
8. Siegbahn, M., et al. (Eds.), (1967). *Atomic, Molecular and Solid State Structure Studied by Means of Electron Spectroscopy*, Almquist: Uppsala.
9. Heilbron, J. L. (1974). *H. G. J. Moseley, The Life and Letters of an English Physicist 1887–1915*, University of California Press.

CHAPTER

2

INDUSTRIAL APPLICATIONS OF X-RAYS

2.1 INTRODUCTION

The industrial uses of X-ray based techniques are many and varied. As was described in chapter 1, the three basic properties of X-ray photons are scattering, photoelectric absorption, and fluorescence. Each of these three properties has been utilized for different industrial applications [1]. Figure 2-1 illustrates eight such applications. The first of these is the X-ray fluorescence technique, in which the wavelengths (figure 2-1a) or energies (figure 2-1b) of fluorescence emission lines are measured to establish the elemental composition of a sample. Use of the intensities of the wavelengths can also allow quantitation of each identified element. Chapter 1 also discussed the possibility of using the X-ray powder diffraction pattern as an empirical "fingerprint" to establish which compounds (phases) make up a specimen (figure 2-1c). The powder method is generally considered to be an empirical method, because the calculations required to decide which lines arise from which set(s) of interplanar spacings can be rather complicated, due mainly to the superposition of lines and uncertainties about the line intensities.

A second important application of the X-ray diffraction method is the *single crystal method* (figure 2-1d). In this method, a single crystal is fixed in a specific orientation relative to the X-ray beam, and diffraction maxima sought. The orientation of the crystal is then changed, and diffraction maxima again sought. This process is repeated as many times as is necessary to build up a detailed picture of the intensity distribution of diffracted radiation coming from the specimen. The effect of repeating many measurements at different crystal orientations is to reduce the rather complicated

19

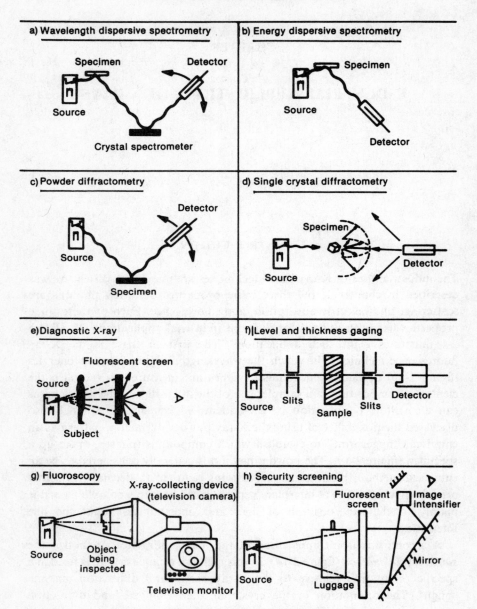

Figure 2-1. Uses of X-ray methods in industry and research.

three dimensional lattice calculations to a series of much simpler calculations. Thus, from single crystal diffraction data it is possible to establish very detailed information about the structure of the specimen. Even though the calculations are greatly simplified, they are still rather complicated, and need all of the power of a modern computer for their solution. This, and the fact that the technique is only applicable to pure phases from which small single crystals can be grown, make it a specialty of the research laboratory, rather than giving it much direct application in industry.

The other four examples given in figure 2-1 are all based on the absorption of X-rays. In the following sections each of these last four methods will be discussed in detail. Briefly, the diagnostic use of X-rays (figure 2-1e) is based on the fact that X-ray photons are more strongly absorbed by high atomic numbers as they are by low atomic numbers. Hence, if a human limb is placed between the X-ray source and a film, a shadowgraph of the bone structure will be observed. The absorption technique can also be used for measuring thickness and the depth of absorbing liquids in a opaque container, as in level and thickness gauging figure (2-1f). Where the technique is used for the examination of the gross structure of metal and similar objects, it is referred to as X-ray fluoroscopy (figure 2-1g). In the last technique, security screening (figure 2-1h), X-rays are used to evaluate the contents of luggage.

2.2 DIAGNOSTIC USES OF X-RAYS

One of the first practical applications of the use of X-rays was for the radiographic analysis of the human body. Roentgen's original paper, published in 1898 [2], included a photograph of his wife's hand clearly showing the bone structure of the fingers. The potential of this technique was quickly realized and methods had already become very sophisticated by the early 1900s. It has been estimated that in the 1914–1918 war as many lives were saved by the use of radiographic techniques as were actually lost in the conflict. The radiographic technique depends on the fact that the absorption of X-rays varies as the third power of atomic number; hence, in general, higher atomic number elements will absorb X-ray photons of a given wavelength more than will lower atomic number elements. Since the bone structure of the human body is high in calcium ($Z = 20$), the bones will absorb X-rays to a greater extent than will the flesh and soft tissue; thus if a limb is placed between a polychromatic beam of X-rays and a piece of photographic film, transmitted photons falling onto the film will produce a shadowgraph of the bone structure of the limb. In the design of systems for radiographic analysis one of the variables is the response characteristics of the film, this in turn being dependent upon the graininess of the emulsion. Both contrast and sharpness of the film image quality are important to the radiologist, who is invariably seeking to observe minor flaws and discontinuities in the image. Exposure time and image quality can be improved by the use of a sandwich arrangement of double-sided emulsion between phosphor screens. A major disadvantage of film is that it represents a passive rather than an active detection device. If a given exposure fails to reveal the detail sought by the radiographer, the exposure may have to be repeated several times.

By direct use of a fluorescence screen in place of the film it is, in principle, possible to obtain a real time image. However, in practice, a rather intense primary source would be required in order to give an observable signal. Use of solid state amplifiers as an image intensifier overcomes this problem, and today there are a wide variety of these devices available [3]. In its simplest form the solid state amplifier is made up of an X-ray-sensitive photocathode plus an electroluminescent screen. The photoconductor changes the local field strength of the electroluminescent screen as a function of X-ray flux, producing light and dark areas. Figure 2-2 shows a schematic view of a modern image intensifier in which primary electrons produced when X-ray photons fall onto the photosensitive layer are accelerated and focussed onto the output screen. A closed circuit TV system is typically coupled to the output port of the image intensifier for fluoroscopic examination.

An important development in real time imaging was the vidicon tube [4]. The term "vidicon" was first applied to the tubes that replaced the combination of photoelectric charge storage and TV tube, the development of which opened up the way towards real time imaging. The original vidicons used a Sb_2S_3 photoconductor that was scanned over its rear surface by an electron beam. As the beam scans across the surface of the photoconductor, it deposits a small negative charge bringing the scanned area to the same potential as the cathode. X-ray photons falling onto the front surface induce conductivity, producing a positive potential proportional to the quantity of radiation. When the scanning beam returns to this particular area, a certain negative charge is required to return the area to cathode potential, and it is the deposition of this negative charge that is used to generate a display signal. Most modern vidicon systems use PbO, or amorphous Se, as the photoconductive surface.

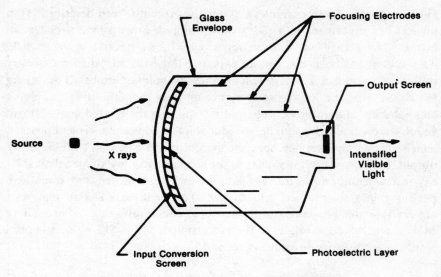

Figure 2-2. Conventional X-ray image intensifier tube.

Conventional real time fluoroscopy uses a vidicon type detector and allows viewing of the image on a TV monitor screen. Use of such a system enables a real time visualization of dynamically changing situations within the living body. A major experimental problem always remains the obtaining of sufficient contrast between the volume of interest and the immediate surroundings. Unfortunately, blood and soft tissue have very similar absorption characteristics, since their average atomic numbers are almost identical. Contrast can sometimes be enhanced by ingestion or injection of contrast agents immediately before or even during fluoroscopic examination. A simple example is the ingestion of a "barium meal" by a patient before fluoroscopic examination of the digestive system. In arteriography, injection of image enhancers into the blood stream is employed.

Another method of image improvement is the use of data processing techniques such as image subtraction or image enhancement. In the former method the significant intrinsic noise associated with the vidicon–TV tube system can be almost completely removed by first storing the experimental image in a suitable digitized form and then subtracting a similar digitized stored image representing the average dark field pixel pattern. Use of the lower noise level Reticon systems also helps to reduce the overall background level, this system having about a sixfold better signal/noise ratio than the vidicon system. As an example, a system has recently been described [5] in which a reticon camera with a 32×32 photodiode array is used to collect an image, which is then fed via an analog to digital converter to a magnetic disk. This system has been used on an experimental basis to detect pulmonary pulsations in monkeys. Image enhancement techniques are based on taking the signal from the X-ray detector, storing it as a digital image, applying computer based data processing, and redisplaying the treated image. It can be looked on as a high speed noise-reduction and edge enhancement process, which acts as an interface between the detector system and the final TV output. In many ways it is similar to the sequence employed in analytical X-ray instrumentation, where raw data are collected, stored and smoothed, peak maxima sought and the treated data redisplayed. For real time applications this processing of fluoroscopic data allows rates of around 10 MHz pixels, which translates to the processing of $512 \times 512 = 262,144$ pixel words in a TV frame in about $\frac{1}{30}$ second.

2.3   TOMOGRAPHY

One of the newest and most valuable techniques of radiography is the method of tomography. In normal radiography one seeks to keep the image as sharp as possible during an exposure, and to this end, relative movement of source, object and detection medium during exposure should be kept at an absolute minimum. In tomography, suitable relative movement is introduced in all planes of the object except for the one of interest. This renders all undesirable planes blurred except for the plane of interest, which is kept sharp. Figure 2-3 compares normal projective imaging with tomographic imaging. In the former case the two orthogonally placed objects produce an image of similar sharpness. In the latter case movement of the detector during the course of a contiguous series of exposures focuses just one of the orthogonal members. Tomography could thus be described as sectional roentgenography. By giving the X-ray tube a curvilinear motion during exposure, but synchronized with the recording medium in the opposite direction, the shadow of the selected plane remains stationary on the film while all other planes have a relative displacement and appear blurred or obliterated. This gives a three-dimensional effect to the viewed image. The tomographic method is invaluable to the radiographer in that it allows revelation of three dimensional detail that would be completely masked by conventional fluoroscopy.

The actual hardware for tomographic work typically allows the perform-

Projective Imaging

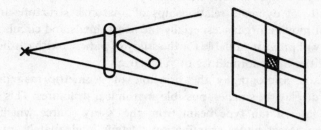

Tomographic Imaging

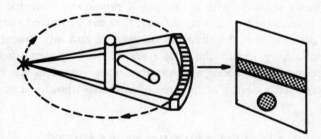

Figure 2-3. X-ray fluoroscopy by projective and tomographic imaging.

ing of multiple exposures. Traditionally this is done using a single X-ray source and a multiple array detector. As an example, the Philips Tomoscan 310 uses a rotating anode X-ray tube that runs at 280 kW, generating a pulse width of 2 msec and a pulse frequency of 8 msec. The detector is a xenon gas ionization chamber with 576 individual detector elements. This gives excellent quality pictures in extremely short exposure times—typically of the order of a few seconds. Very sophisticated computer software is also available with these systems, allowing great flexibility in pattern storage, enhancement, redisplay of areas of interest and so on.

Newer developments in tomographic instrumentation utilize multiple X-rays sources allowing an even higher degree of flexibility. One such sytem [6] consists of twenty-eight X-ray sources arranged in a semicircle with twenty-eight corresponding image intensifiers. This entire system permits mathematical reconstruction imaging of a cylindrical three-dimensional volume of 23 cm axially and 30 cm diameter transaxially. This system should be applicable to the study of dynamic relationships of anatomic structure and to the function of moving organs, especially the heart, lungs and circulation. This technique will prove invaluable for the study of patients with cardiopulmonary disabilities or abnormalities of vascular anatomy.

In vascular radiography the principle of geometric magnification is applied to display the smallest possible anatomical structures. This technique essentially uses a fan type beam from the X-ray source which diverges through the object under examination. Details inadequately imaged in a normal radiograph become strikingly visible using this approach, making comprehensive and accurate diagnoses much more probable. The same basic idea can be applied to tomography, but with even greater advantage in that the inefficient use of detectors generally encountered in conventional fan beam geometry is avoided by coupling together the source and detector. A direct fan beam scanner, with a rotating X-ray source–detector assembly attached to it, is mounted on a rigid frame that can be moved radially with respect to the isocenter. Mechanical movements and adjustments can be made automatically within the whole gantry without repositioning the patient. Reconstructive zoom and variable image regeneration capability provides increased visibility of the area of interest without rescanning.

2.4   LEVEL AND THICKNESS GAUGING

The thickness of a solid homogeneous material is readily measurable by X-ray absorptiometry, and this method can be applied both for the estimation of bulk thickness and for the measurement of layer thickness. Figure 2-4 shows the three basic techniques that are employed for X-ray gauging and indicates a specimen being irradiated by a collimated primary beam of X-ray photons that may, or may not, have been filtered to change the spectral distribution. In the backscattering arrangement (A), a collimated detector with an optional filter is placed at an angle from the incident beam but on the same side of the specimen as that beam. Both scattered and fluorescent radiation from the specimen can now enter the detector. In the fluorescence arrangement (B), a spectrometer is interposed between the specimen and the detector such that only one or more selected wavelengths can enter the detector. Finally, in the transmission arrangement (C) a collimated detector is placed on the opposite side of the specimen from the source, so that only radiation transmitted by the specimen can enter the detector.

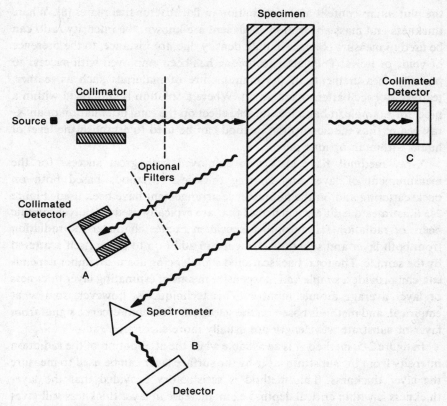

Figure 2-4. Three basic methods for X-ray thickness gauging: (*A*) backscattering position; (*B*) fluorescence position; (*C*) transmission position.

The most common method of measuring the thickness or consistency of a material is using the transmission mode (C). As will be seen from equation (1-3), the attenuation of the X-ray beam is dependent upon specimen density, thickness and mass absorption coefficient. Thus for a given material, e.g. steel, where the density and absorption coefficient are known, the ratio of incident to transmitted beam intensity can be used to measure the thickness of the steel. As an example, completely automatic measurement of the thickness of sheet steel using this technique has been performed for many years [7]. Alternatively, where the apparent thickness is known, the same intensity ratio can be used to monitor the variations in absorption coefficient (and therefore average atomic number) of the material contained within this layer. As an example, this technique has been used for the determination of the plutonium content and distribution in flat reactor fuel plates [8]. Where thickness and mass absorption coefficient are known, the intensity ratio can be used to measure changes in bulk density due, for instance, to the presence of voids or holes. This latter technique has been employed with success to porosity measurements in the manufacture of materials such as leather, textiles, storage batteries and so on. Where a solution is contained within a pipe, for example, it has a measurable effect on the total attenuation of an X-ray beam; thus the absorption method can be used to establish the level of liquid within an opaque pipe.

X-ray methods have also been employed with great success for the measurement of layer and coating thickness. Methods based both on backscattering and on the fluorescence arrangement have been used. Figure 2-5 illustrates the different schemes that are typically used. A polychromatic beam of radiation falling onto a specimen excites characteristic radiation from both layer and substrate elements, in addition to being itself scattered by the sample. The total backscattered signal, being atomic number dependent, can provide a simple and inexpensive means of estimating layer thickness or layer average atomic number. This technique is, however, somewhat empirical, and methods based on the selection of the fluorescence signal from layer or substrate wavelength are usually more successful.

In figure 2-5, method A is applicable when the attenuation of the radiation intensity from the substrate layer by the surface layer can be used to measure the layer thickness. This method is satisfactory provided that the layer thickness is within critical depth, i.e., an increase in layer thickness still gives a measurable change in the transmitted intensity. Method B can be used where a single wavelength can be selected from the substrate layer radiation. By the use of filters or a spectrometer channel, the calculation of the layer thickness becomes much easier and much more precise, since only one value of the mass absorption coefficient has to be considered. Method C is used where the layer thickness is within critical depth of the measured wavelength. In this instance, the intensity of characteristic radiation from a layer element can be used to estimate the layer thickness.

2.5 NONDESTRUCTIVE TESTING

X-ray imaging tests are among the methods most widely used to examine interior regions of metal castings, fusion weldments, composite structures and brazed components. Radiographic tests are made on pipeline welds,

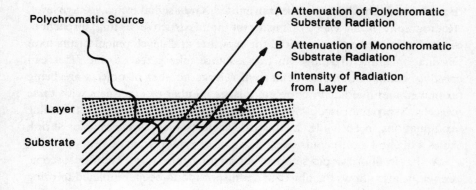

Polychromatic Source

A Attenuation of Polychromatic Substrate Radiation

B Attenuation of Monochromatic Substrate Radiation

C Intensity of Radiation from Layer

Layer

Substrate

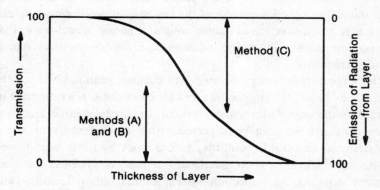

Figure 2-5. Methods for the measurement of coating thickness.

pressure vessels, nuclear fuel rods and other critical materials and components that may contain three dimensional voids, inclusions, gaps or cracks aligned so that the critical areas are parallel to the X-ray beam. Since penetrating radiation tests depend upon the absorption properties of materials for X-ray photons, the tests can reveal changes in thickness and density and to the presence of inclusions in the material under examination.

Essentially two basic techniques are employed. In X-ray radiography the sample being examined is placed between an X-ray source and a film, and an exposure will produce an absorption shadowgraph of the object being irradiated. In X-ray fluoroscopy the film is replaced by some form of X-ray transducer that will convert the transmitted X-ray signal into a voltage level. Radiography is an effective means of nondestructive testing inspection, limited only by the time required for exposure and development of the film. Evaluation must, of course, in this instance take place off line. This can produce complications, particularly if a large number of objects are being examined and one has to correlate a large number of exposures with these objects. X-ray fluoroscopy avoids this problem and allows direct, on-line examinations to be made. In this case, defects such as porosity or shrink holes will show up as contrasting areas on the fluorescent screen.

A simple fluoroscopic setup is shown in figure 2-1g. The fluoroscopic technique also allows the object under inspection to be manipulated to bring it into an optimum viewing position. It can thus be observed in motion on a screen in front of the operator, providing a facility for virtual three dimensional observation of the size and shape of a detect or inclusion. In most commercially available systems the object on the screen can be enlarged, thus allowing optimization of the image quality. Fluoroscopy can be integrated into a production or quality control situation in such a way that each inspection takes only a few minutes and no intermediate file of paperwork is required.

For the inspection of heavy or awkward castings manipulation of the object itself may be virtually impossible. In such cases the X-ray tube must be transported to the object and maneuvered to obtain the desired angle for inspection. Castings that can be inspected within an enclosed area, e.g. a specially designed lead-lined room, may be examined by tubes mounted on mobile or fixed stands, or on overhead suspension systems. Such an arrangement provides an inspection system of maximum flexibility and accessibility that is capable of handling diverse castings of almost any shape and size.

In recent years, technological advances in fluoroscopy have drastically reduced inspection and interpretation times, while providing a level of picture quality and resolution able to meet extremely rigid specifications for light alloy and iron or steel castings, at a thickness used in the majority of cast components. At the same time, the incentive to tighten inspection criteria for castings has increased. As an example, the drive to conserve energy in the automobile industry both during and after manufacture has led to the use of more lightweight alloys and new design concepts.

In many areas of modern day research it is necessary to establish what happens inside a given sample, and fluoroscopic techniques allow such examinations to be made—for instance, in the examination of a nuclear fuel rod container, or discovering how a projectile behaves inside a gun barrel, and so on.

In the food processing and packaging industry X-ray inspection is becoming increasingly important. With growing output and greater use of mass production, the chances increase of accidentially introducing foreign bodies, such as glass or metallic particles.

2.6 SECURITY SCREENING SYSTEMS

Within a year of the discovery of X-rays by Roentgen, X-ray imaging was being employed not just for medical applications but also by customs and security personnel for examining the contents of packages. As an example, at the turn of this century, X-ray security screening was used in Paris for detecting the contents of parcels sent to the French Chamber of Deputies. Post office authorities used them to detect coins that were mailed hidden in newspaper or sealing wax, contrary to the regulations of the time. Some attempts were also made to apply the technique in the areas of zoology and paleontology. Objects including ancient mummies and carcasses of sacred Egyptian ibises were examined at the Vienna Museum of Natural History.

Although the full potential of the penetrating power of X-rays was quickly appreciated, its implementation at that time was rather impractical and hence proved of minimal value to use in security screening, in contrast to its use by the medical profession. The use of X-rays as a basis for security systems was very limited for a period of almost 70 years even though manufacturers were offering industrial or medical units for security use. These machines were primarily designed for other functions, and the units failed to develop a market; no more than a few dozen units were ever actually used for security screening.

However, there was a renewal of interest in the use of X-ray methods for security screening in the early 1960s, mainly for bomb disposal work, and for the screening of mail of prominent people. The need for large dose rates and consequent safety hazards required the use of rather massive units that were both bulky and slow in operation. Nearly all of this early equipment was still primarily based on designs originally intended for medical and industrial use, and they comprised essentially a cabinet unit fitted with a fluorescent screen viewed through lead glass. X-ray machines finally became viable for security applications in the late 1970s [9], the most important development being the ability to generate an X-ray image with a very low dose rate of X-rays, so low in fact that the radiation would not even damage photographic film.

The requirements of security X-ray systems generally demand that they be enclosed, self-contained units (called cabinet X-ray systems) which can be operated by relatively unskilled personnel. The units must also be absolutely radiation safe under all operating conditions. A low dose rate unit employs an X-ray dose rate something like 50,000 times less, in terms of radiation flux, than in high dose systems with direct viewing fluorescent screens. These requirements can be realized in hardware by putting the X-ray system in a cabinet, with specifications depending on the degree of inspection detail wanted, the inspection rate required, the maximum parcel size to be examined, and the physical layout of the actual hardware. Low dose systems are usually designed around a radiation shielded cabinet in which the radiation source itself is contained within radiation safe enclosure. A conventional system has a control panel, an X-ray generator, a viewing system, and a door or port through which the luggage to be inspected is loaded. Because low dose systems require much less shielding than high dose systems, items to be inspected can be transported through them at high speed via flexible lead curtains that act as baffles.

A typical unit is illustrated in figure 2-6. The system operates continuously with a low level X-ray beam which, when on, causes a fluorescent screen to illuminate. The screen is viewed by a multistage light amplifier that is attached to a closed circuit television camera for viewing on a TV monitor, or an output lens for direct viewing of the phosphor of the last light amplifier module. The pulsed X-ray system produces a single X-ray pulse which momentarily causes the screen to be illuminated. During this time a closed circuit TV camera views the screen and the video signal is stored. The image is then repeatedly displayed on a TV monitor. The storage device must keep refreshing the TV monitor picture and is cleared before the next inspection cycle. The scanning X-ray beam system operates on a two dimensional raster scan in which the beam moves in a vertical line while a conveyor belt moves the parcel under examination in the horizontal plane. By chopping the vertical beam and storing it in an x/y raster that is displayed on the TV monitor, the X-ray image is presented as a series of closely spaced dots of differing brightness. The storage device keeps refreshing the TV monitor picture in the same manner as the pulsed X-ray system.

By the early 1980s, there were approximately 1500 low dose systems in the United States, mainly at transportation and public sites. In addition, there were around 200–300 high dose systems at governmental and private security

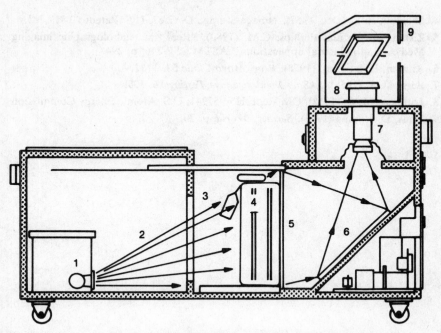

Figure 2-6. A low dose X-ray security screening system: (1) X-ray tube, (2) X-ray beam, (3) X-ray chamber, (4) item being inspected, (5) fluorescent screen, (6) mirror, (7) light amplifier or television camera, (8) adjustable viewing mirror, (9) adjustable viewing hood.

locations. All of the low dose systems are of the "film safe" type, since a high percentage of the travelling public carry film and cameras on their persons. In the U.S.A. several thousand weapons and other prohibited articles are detected by these screening systems each year. In addition to their actual ability to detect threat items, they also offer a tremendous psychological deterrence to both amateur and professional criminals.

REFERENCES

1. Jenkins, R. (1984). "X-ray technology" *In* Kirk and Othmer (Eds.), *Encyclopedia of Chemical Technology*, Vol. 24, Wiley: New York, pp. 678–708.
2. Roentgen, W. C. (1898). *Ann. Phys. Chem.* **64**, 1.
3. Kuhl, W. (1980). "Real time radiographic imaging: Medical and industrial applications," ASTM STP-716, p. 33.

4. Sheldon, E. E. (Dec. 1957). "Image Storage Device", U.S. Patent #2,817,781.

5. Ziskin, M. C., and Phillips, C.M. (1980). "Real time radiolographic imaging: Medical and industrial applications," ASTM STP-716, p. 294.

6. Ritman, E. L., et al. (1978). *Proc. Mayo Clinic* **53**, 3–11.

7. Berstein, S. (1958). *J. Soc. Nondestructive Testing* **16**, 305.

8. Lambert, M. C. (1958). OM Rept. HW-57941, U.S. Atomic Energy Commission.

9. Haas, D. J. (Aug. 1976). *Security World*, p. 20.

CHAPTER

3

X-RAY DIFFRACTION

3.1 USE OF X-RAY DIFFRACTION TO STUDY THE CRYSTALLINE STATE

A crystal consists of atoms or molecules arranged in patterns that are repeated regularly in three dimensions. The smallest repeat unit, the "unit cell," of this three-dimensional unit may consist of one or many atoms. In a given crystal all of the individual unit cells must be identical in orientation and composition. While the term *crystallinity* is commonly used to explain diffraction phenomena, it is not an ideal term, since it is generally related to physical characteristics, such as shape and luster, observable by eye or with the optical microscope. In practice, there are many cases of materials giving diffraction patterns, where such physical properties are not visibly apparent, because the size of the particles is so small. In fact, the property that gives rise to the interference of the coherently scattered X-rays, and hence to the diffraction pattern, is *order*. Almost all solid materials exhibit some degree of regular order, and (as was shown in section 1.6) under certain experimental circumstances this order will give rise to an X-ray diffraction pattern. The X-ray pattern is characteristic of the material whence it was derived, because each compound is made up of a similarly unique combination and arrangement of atoms. The X-ray diffraction pattern can thus be used to characterize materials, and this is the basis of the X-ray powder method.

X-ray patterns are recorded with an (almost) monochromatic X-ray source, and each diffraction peak angle corresponds to one or more *d*-spacings. Bragg's law, equation (1-8), is used to convert each observed peak maximum measured in degrees (2θ) to *d*-spacing. When using Bragg's law for powder diffraction, it is customary to write it in a form in which n does not appear explicitly:

$$\lambda = \frac{2d}{n} \sin \theta \qquad (3\text{-}1)$$

The factor d/n is the d-value, which is almost invariably used in powder diffraction. It should be noted that d/n is a submultiple of the "true" separation between adjacent planes that pass through lattice points. The value of λ is known to a few parts per million; thus measurements of angles at which diffraction occurs make possible the determination of the d-values with relative ease and with high precision. It will also be seen by looking at the Bragg law that since the maximum value of θ is 90°, and hence the maximum value of $\sin \theta$ is unity, the minimum detectable value of d will be equal to $\lambda/2$. The maximum value of d is not limited in any fundamental way but is generally determined by the experimental arrangement.

Useful additional information can be derived by differentiating Bragg's law:

$$\Delta\lambda = 2d \cos \theta \, \Delta\theta + 2 \, \Delta d \sin \theta \qquad (3\text{-}2)$$

Since λ is a constant, $\Delta\lambda$ equals zero, giving

$$\frac{\Delta d}{d} = -\cos \theta \, \Delta\theta \qquad (3\text{-}3)$$

Equation 3-3 is a very useful measure of the relative error in d that results from a given error in θ, that is from $\Delta\theta$. The latter must be expressed in radians. As θ approaches 90°, $\cot \theta$ approaches zero. At 90°, therefore, the value of $\Delta d/d$ will always be zero regardless of the error of angular measurement. It is evident that the most precise results will be made at angles as close to 90° as possible.

3.2 THE POWDER METHOD

Of all of the methods available to the analytical chemist for materials characterization, only X-ray diffraction is capable of providing general purpose qualitative and quantitative information about the presence of phases (e.g. compounds) in an unknown mixture. While it is true that techniques such as differential thermal analysis will provide some information on specific phase systems, such methods cannot be classified as general purpose. As described in the previous section, a diffraction pattern is characteristic of the atomic arrangement within a given phase, and to this extent it acts as a "fingerprint" of that particular phase. The powder method derives its name from the fact that the specimen is typically in the form of a microcrystalline powder, although, as has been previously indicated, any material which is made up of an ordered array of atoms will give a diffraction pattern. The possibility of using a diffraction pattern as a means of phase identification was recognized many years ago, but it was not until the late 1930s, that a systematic means of unscrambling, or "search–matching" the superimposed diffraction patterns was proposed [1, 2]. The search–matching technique is based on the use of a file, called the Powder Diffraction File (PDF), which consists of a collection of single phase reference patterns, characterized in the first stage by their strongest reflections. The search technique is based on finding potential matches by comparing the strong lines in the unknown pattern with these standard pattern lines. A potential match is then confirmed by a check using the full pattern in question. The identified pattern is then subtracted from the experimental pattern and the procedure repeated on the residue pattern until all lines are identified.

Techniques for this search–matching process have changed little over the years, and although in the hands of experts manual search–matching is an extremely powerful tool, for the less experienced user it can be rather time consuming. The responsibility for the maintenance of the Powder Diffraction File lies with the International Centre for Diffraction Data (ICDD), Swarthmore, Pennsylvania, U.S.A. This group is made up of a staff of permanent officers along with a number of academic and industrial scientists who are active in the field of X-ray powder diffractometry. The Powder Diffraction File is a unique assembly of good quality single phase patterns and is used by thousands of chemists, geologists, materials scientists, etc., all over the world. In recent years, the automation of the search–matching process has made the routine use of the method even more widespread by doing much to relieve the tedium associated with manual search–matching.

The two basic parameters used in the search–matching process are the d-values, which have in turn been calculated from the measured 2θ values and the relative intensities of the lines. Whereas the d-value can be precisely measured—perhaps with an accuracy of better than 0.5% in routine analysis—the measured intensities are rather unreliable and can be subject to large errors, due in turn mainly to orientation problems. The ease of any qualitative search procedure is also greatly influenced by the quality of the standard d/I values employed for comparison. The quality of the data in the PDF is quite variable and is probably of the order of 1–20 parts per thousand in d, and 5–50% in intensity. Because of the great uncertainty in experimental intensity values, most search procedures give high credence to the d-values but much less to the intensities. It is important, therefore, that the greatest care be taken in the estimation of experimental d-values.

3.3   USE OF X-RAY POWDER CAMERAS

The measured parameters in X-ray powder diffraction are the 2θ maxima and their relative intensities. Each 2θ value is converted to a d-spacing using Bragg's law, to generate a list of d-spacings and intensities called a d/I *list*. There are two basic classes of instruments used for the measurement of a diffraction pattern: diffractometers and film cameras (see figure 3-1). While most modern laboratories employ diffractometers for routine work, the somewhat slower camera methods offer certain advantages and still find useful application. Two camera methods are in fairly common use: the Debye–Scherrer method and the Guinier method. Many factors can contribute to the decision to select a camera technique rather than the powder diffractometer for obtaining experimental data, including lack of adequate sample quantity, the need to investigate sample texture, the desire for maximum resolution, and, possibly most important, the lower initial cost.

The upper portion of figure 3-1 illustrates a typical Debye–Scherrer camera. The component parts include the light tight camera body and cover that hold the sample mount SP and film FM; a secondary collimator SC, which also acts as the main beam stop; and an incident beam collimator PC. A thin metal filter FR is placed between the X-ray source S and the entrance collimator, to remove the unwanted Kβ component plus some continuum from the incident beam, while passing most of the Kα. During operation, the incident beam collimator directs a narrow beam of X-rays onto the sample, which is mounted at the center of the camera. A second collimator is generally placed behind the specimen to prevent scattered radiation from spoiling the film. The sample is packed in a lithium glass capillary or mounted on the tip of a glass fibre. The sample on its mount is placed at the camera center and aligned to the main beam before film loading, by viewing through the collimator or an alignment microscope. Since X-ray diffraction is

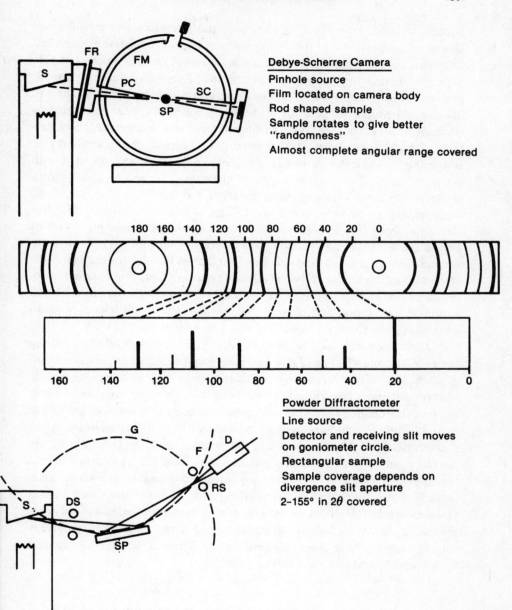

Debye-Scherrer Camera

Pinhole source

Film located on camera body

Rod shaped sample

Sample rotates to give better "randomness"

Almost complete angular range covered

Powder Diffractometer

Line source

Detector and receiving slit moves on goniometer circle.

Rectangular sample

Sample coverage depends on divergence slit aperture

2–155° in 2θ covered

Figure 3-1. Measurement of X-ray diffraction patterns.

a bulk or mass dependent phenomenon, the intensities of the diffracted beams are proportional to the sample volume. Exposure times are dependent on the size and crystallinity of the sample. An ideal Debye–Scherrer sample is polycrystalline and is composed thousands of crystallites around 5 to 10 μm in size. The sample is rotated during exposure, causing the crystallites to assume all possible orientations in the X-ray beam. A crystallite is oriented properly to the beam for diffraction when the incident beam makes an angle θ with a set of lattice planes. At this instant, a diffracted beam is emitted that also makes an angle θ with the lattice planes and therefore an angle of 2θ with the main beam. If the sample is ground fine enough, a smooth and uniform circle of diffracted intensity will appear. If the particle size is too large, however, the circle will appear spotty and possibly incomplete. The line intensities from Debye–Scherrer films are normally estimated on a scale of one to ten by visual comparison with a series of lines of varying intensity registered on a standard exposed film. Two factors that contribute to the low resolution of the Debye–Scherrer camera types are first, that incident beam divergence still exists in a collimator designed to yield the narrowest beam with usable intensity, and second, that the X-ray filter still passes a broad wavelength band that includes $K\alpha_1$ and $K\alpha_2$ radiation.

Higher resolution can be obtained by going to the Guinier focusing geometry camera [3]. The Guinier geometry makes use of an incident beam monochromator that is cut and oriented to diffract the $K\alpha_1$ component of the incident radiation. In addition to diffracting a very narrow wavelength band, the monochromator curvature is designed to use the whole surface of the crystal to diffract simultaneously, thus yielding a large diffracted intensity. As a consequence of this curvature and the orientation of the crystal itself, the monochromator not only easily separates $K\alpha_1$ from $K\alpha_2$, but also converts the divergent incident beam into an intense convergent diffracted beam focused onto a sharp line. Although the quality of data from a well-aligned Guinier camera is high, the camera does have its limitations, especially in the estimation of line intensities. A well-equipped X-ray diffraction laboratory should probably have access to both Debye–Scherrer and Guinier camera techniques, in addition to a diffractometer.

3.4   THE POWDER DIFFRACTOMETER

The lower portion of figure 3-1 illustrates the powder diffractometer. The powder diffractometer was developed in the late 1940s and has changed little from the original concept. The major difference that is found in modern instrumentation is the use of the minicomputer for control, data acquisition and data processing. The geometric arrangement employed in the powder diffractometer is known as the Bragg–Brentano para-focussing system and is typified by a diverging beam from an X-ray line source S, falling onto a large flat specimen SP, being diffracted and passing through a receiving slit RS to the detector D. The amount of divergence is determined by the effective focal width of the source and the aperture of the divergence slit. Use of the narrower divergence slit will give a smaller specimen coverage at a given diffraction angle, thus allowing the attainment of lower diffraction angles where the specimen has a larger apparent surface (thus larger values of d are attainable). Axial divergence is controlled by two sets of parallel plate collimators (Soller slits) placed between focus and specimen and between specimen and scatter slit respectively. Two circles are generated by this geometry—the focusing circle F, and the goniometer circle G. The source, the specimen and the receiving slit all lie on the focusing circle, which has a variable radius. The specimen lies at the center of the goniometer circle, which has a fixed radius. The instrumental line width of the diffracted profile will be determined mainly by the angular aperture of the receiving slit, but the intensity of this line will be dependent both on the slit aperture and on the focal spot characteristics of the X-ray tube.

A photon detector, typically a scintillation detector, is placed behind the receiving slit, and this converts the diffracted X-ray photons into voltage pulses. These pulses may be integrated in a rate meter to give an analog signal on an x–t recorder. By synchronizing the scanning speed of the goniometer with the recorder, a plot is obtained of 2θ vs intensity. This plot is called a diffractogram. A timer–scaler is also provided for quantitative work; it is used to obtain a measure of the integrated peak intensity of one or more selected lines from each analyte phase in the specimen. A diffracted beam monochromator may also be used in order to improve the signal to noise characteristics. The output from the diffractometer is a powder diagram, essentially a plot of intensity as a function of diffraction angle, which may be in the form of a strip chart or hard copy from a computer graphics terminal.

3.5 QUALITATIVE APPLICATIONS OF THE X-RAY POWDER METHOD

As has been previously stated, a polycrystalline phase diffracts a monochromatic beam of X-rays into a spectrum, or *pattern*, of lines that can be recorded by diffractometer or camera techniques. Each line in the diffraction pattern, in terms of its position and intensity, represents a particular family of crystal planes in the phase. Since the sequence of planes is unique to the phase, the X-ray pattern is itself specific enough to act as a "fingerprint" for the identification of the phase. The patterns of individual phases differ widely in the sequence and intensities of their lines and these effects are used in the qualitative identification process generally known as search–matching. A search manual is used for the identification of any phase whose pattern is in the manual. The complete PDF comprises nearly 50,000 phases in 1988. A simple alphabetical index based upon the name or the chemical formula of a material provides one practical means of very effectively using this large reference library of patterns. Because of the problems of pattern recognition with a data file this large, the PDF is also available as subsets. Currently available are subfiles of inorganics, organics, minerals, metals and alloys, NBS patterns, forensics and common phases; and possible new subfiles are always under consideration. These data are available on cards, on micro-fiches, in books, and on computer disks, magnetic tapes and CD ROM. Concurrently with the buildup of the PDF, there has been considerable effort to develop both manual and computer-based retrieval methods, to match an unknown pattern to one in the PDF. The retrieval of patterns is somewhat analogous to the matching of human fingerprints, which makes use of the characteristic loops, whorls, arches, pockets, etc. The problem of pattern matching in X-ray powder diffraction is generally easy for patterns obtained from single phases, but becomes more complicated for actual unknowns where the pattern may represent a mixture of two or more phases.

The most commonly employed manual search–matching procedure is the Hanawalt system. The Hanawalt search manual is organized into groups and subgroups. Two lines from the table of d-values serve to locate the entry of the pattern in the manual. The d-value of the strongest line of the pattern determines the entry group, sometimes called *the Hanawalt group*, and the second strongest line determines the subgroup, that is, the location within that group. An individual entry also lists, in order of decreasing intensities, the d-values of the next six strongest lines in the pattern. In order to identify an unknown, the first and second strongest lines are chosen as the line pair with which to look for a matching entry in the search manual. If the remaining six lines of a qualifying entry also match lines in the unknown pattern, then the identification is can be considered certain. The final step in the procedure is to confirm the identification by going to the PDF to check the complete standard reference pattern.

If the choice of line-pair does not lead to identification, one proceeds by choosing successively weaker lines to use with first line. If still unsuccessful, a new start is made with a different choice of first line.

Although the Hanawalt method for searching the powder data file is the most popular search–matching method, it has not always proven completely successful for identifying unknown diffraction patterns in which the intensity values are markedly different from those in the file patterns, even though the d-values themselves might be very accurate. The problem may arise when fewer than two of the three strong lines in the experimental pattern are recognized for use in the search procedure. This situation often occurs when data are obtained by electron or neutron diffraction. In the X-ray diffraction case it may also occur when the sample shows a very strong preferred orientation, or where there are insufficient crystallites in the sample to achieve ideal randomness. A alternative search method, which partially compensates for such intensity problems, is the Fink method [4]. The Fink method relies almost exclusively on the largest d-spacings of the pattern. The basic assumption is that, although there may be significant differences in relative intensities in, for example, the electron diffraction pattern, when compared with the corresponding X-ray pattern, most of the strong lines are observable in both patterns. The Fink search manual lists the first eight d-values of each reference pattern. To start the search the user lists the largest d-values from the unknown material and then attempts to find a match based on the d-values.

It will be apparent that a basic sequence of search–matching steps would be amenable to modern computer techniques. In the mid 1960s, three mainframe computer based search programs were described almost simultaneously. In 1965, Frevel [5] described a program to perform an entire search and match process, based on standard files prepared from the PDF. The program was called ZRD, because it uses atomic number Z, functional group information R, and d-spacing data D. Other computer programs were described by Johnson and Vand [6] and by Nichols. In the following years, each of the programs has been modified, either to improve operation, to generalize the application, to include many types of data files and computers, or to incorporate successful features found in other search–matching programs [7, 8]. More recently, most major X-ray instrument manufacturers have developed search–matching programs capable of performing searches on minicomputers, of the size typically supplied with automated powder diffractometers. All are highly dependent on data quality for both the unknown pattern and the standard pattern. In order to quantify the success of computer search–matching and to compare with hand searching techniques, the ICDD has run several round robin tests in which standard samples and data sets have been circulated to typical users for analysis. In each case it was found that high success rates are observed when the unknown and standard reference patterns are of high quality [9, 10].

The rapid incorporation of minicomputers into diffraction instrumentation has provided a great opportunity to perform rapid data collection and sophisticated data reduction, followed by search–matching procedures on an automated or semiautomated basis, completely in house and unattended (e.g. refs. 11, 12). In the United States today more than 90% of all new powder diffractometers sold are automated to some degree, although still less than 10% or so of the 15,000 diffraction users in the world are currently using computers for search–matching. The recently available CD ROM (compact disk read only memory) means that the whole of the Powder Diffraction File can be stored easily along with suitable search–matching programs to be run on a personal computer [13]. This system offers 540 megabytes of storage and is available at reasonable cost. Stand-alone PC based search systems are also now available [14], allowing even the most modestly equipped X-ray laboratory the opportunity to use the very latest automated search–matching software.

3.6 QUANTITATIVE METHODS IN X-RAY POWDER DIFFRACTION

Once the presence of a phase has been established in a given specimen, one can, at least in principle, determine how much of that phase is present, by use of the intensities of one or more diffraction lines from the phase. The intensities of the diffraction peaks are dependent on a number of different factors, falling into three categories:

a. Structure dependent: that is, a function of atomic size and atomic arrangement, plus some dependence on the scattering angle and temperature.

b. Instrument dependent: that is, a function of diffractometer conditions, source power, slit widths, detector efficiency, etc.

c. Specimen dependent: that is, a function of phase composition, specimen absorption, particle size, distribution and orientation.

For a given phase, or selection of phases, all structure dependent terms are fixed and so have no influence on the quantitative procedure. Provided that the diffractometer terms are constant, the instrument dependence can also be ignored. Thus if one calibrates the diffractometer with a sample of the pure phase of interest and then uses the same conditions for the analysis of the same phase in an unknown mixture, only the random errors associated with a given observation of intensity have to be considered. The biggest problem in the quantitative analysis of multiphase mixtures then remains the specimen dependent terms, and specifically those dependent upon particle size and distribution, plus effects of absorption.

The absorption effect has already been referred to in section 1.4, and clearly, in a multiphase mixture, different phases will absorb the diffracted photons by different amounts. As an example, the mass absorption coefficient for Cu Kα radiation is 308 cm^2/g for iron, but only 61 cm^2/g for silicon. Thus iron atoms are five times more efficient in absorbing Cu Kα photons than are silicon atoms. There are a variety of standard procedures for correcting for the absorption problem, of which by far the most common is the use of an internal standard. In this method, a standard phase is chosen which has about the same mass absorption coefficient as the analyte phase. A weighed amount of this material is added to the unknown sample, and the intensities of lines from the analyte phase and from the internal standard phase are then used to estimate the relative concentrations of internal standard and analyte phases. The relative sensitivity of the diffractometer for these two phases is determined by a separate experiment. Other procedures are available for the analysis of complex mixtures, but these are beyond the scope of this book. For further information the reader is referred to texts dealing specifically with the X-ray powder method (e.g. ref. 15).

The treatment of the absorption effect is fairly simple; the handling of particle problems is much more complex. As has been previously stated, the powder method requires a specimen that is randomly oriented, since the geometry of the system requires that an equal number of crystallites be in the correct orientation to diffract at any angle of the goniometer. Where particles lie in a preferred orientation, there will be more particles available to diffract at the angle corresponding to this orientation, and what is equally important, less particles available to diffract at other diffraction angles. The overall effect is illustrated in figure 3-2. Here a beam of radiation from the source S falls onto the sample SP, and is diffracted. The complete diffraction ring is shown, with an even distribution of intensity around it. The diffractometer receiving slit intercepts only a small portion of this ring, and as long as the portion intercepted is representative of the whole ring, there is no problem. However, when the intensity distribution around the ring is

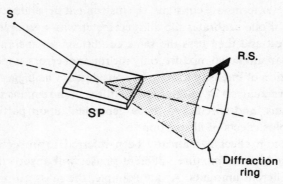

Figure 3-2. Diagram showing how the Bragg–Brentano para-focussing geometry only allows a small portion of the diffraction cone to be detected.

uneven (due, for example, to orientation), the intensity passing through the slit may not be a good measure of the average intensity of the whole ring. The overall effect may be to enhance some intensities and to diminish others. In other words, the intensities are now dependent upon particle distribution and orientation. Some materials, just by virtue of their crystal habit, may become preferred during sample preparation. As an example, mica, being a rather "platy" material, will prefer to stack one plate on top of another rather than take up a random orientation. The overall effect of preferred orientation can vary from insignificance to errors of the order of tens of percent. Careful specimen preparation is always called for, and this may include grinding, sieving, spray drying, or other suitable techniques. Tests have been made comparing these various methods and in extreme cases, as was mentioned previously, the intensities vary by as much as an order of magnitude [16].

The areas of application of quantitative X-ray powder diffraction are many and varied, and hundreds of analysts are using this technique on a daily basis. Some of the more common applications include ore and mineral analysis, quality control of rutile–anatase mixtures, retained austenite in steels, determination of phases in airborne particulates, various thin film applications, study of catalysts and analysis of cements. The current state of the art in the quantitative analysis of multiphase materials is that accuracies of the order of one percent or so can be obtained in those cases where the particle orientation effect is either nonexistent or has been adequately compensated for.

3.7   OTHER APPLICATIONS OF X-RAY DIFFRACTION

In addition to the many applications of qualitative and quantitative phase analysis already described, there are a variety of other applications of X-ray diffraction that should be mentioned. While the diagrams shown in figure 3-3 by no means represent an exhaustive list, they should at least give an indication of the versatility of X-ray diffraction methods. Figure 3-3a illustrates the use of X-ray powder diffractometry at nonambient temperatures. Most major instrument manufacturers supply different attachments to a standard powder diffractometer which allow measurements at temperatures from ambient up to about 2,000°C, or down to the temperature of liquid nitrogen. Such attachments are invaluable for the study of phase transformations, especially where computerized systems allow the display of a three dimensional intensity–2θ angle–temperature diagram (e.g. ref. 17). This technique can be combined with the use of inert or other special atmospheres within the sample chamber. Other applications of high temperature diffraction include the study of thermal expansion, recrystallization, thermal decomposition, solid state addition and replacement reactions, etc.

The second diagram, figure 3-3b, illustrates the use of X-ray diffraction for the study of fibres. In this experimental arrangement, the fibre is mounted in a specific direction between the X-ray source and a flat piece of film. When the exposure is complete and the film is developed, a number of discontinuous diffraction rings will be seen on the film. The lengths, widths and positions of these rings can be used to establish the preferred orientation and the degree of crystallinity of the specimen (see e.g. ref. 15, chapter 10). As was stated in the last section, where nonrandom orientation of particles is present in a specimen, the intensities around a given diffraction ring will be nonuniform. Most natural and artificial fibres show some of this preferred orientation because of the orientation, during growth, of the long chain molecules from which they are made. The physical and mechanical properties of the fibre are very dependent upon this orientation.

The study of multidimensional intensity distribution can also be used for the examination of flat specimens, by representing a stereographic projection of the intensity as a *pole figure* diagram, as illustrated in figure 3-3c. In a pole figure (texture) goniometer, two additional movements are used to rotate and tilt the specimen at a fixed θ/2θ angle setting. Coupling of the tilt and rotation allows the three dimensional space above the specimen to be sampled for X-ray intensity. The best way to conceive of this movement is to imagine a

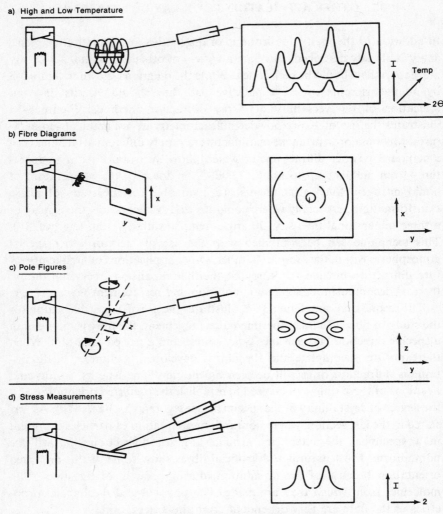

Figure 3-3. Special applications of X-ray diffraction.

continuous line to be drawn from a point on a ball, spiraling away from the point until the line reaches the maximum diameter of the ball. A line normal to the surface of the specimen, and at its center point (the pole), would track this continuous line exactly. Special charts are used to plot the pole figure information. Pole figures are used for the study of preferred orientation—or "texture"—in metals, plastics, thin films, etc., and are invaluable for the study and prediction of flow and plastic deformation due to mechanical treatments, such as rolling and pressing.

Another important use of X-ray diffraction in industrial processes is for the study of residual stress, illustrated in figure 3-3d. When a uniform metal bar is subject to stress, either by tension, compression or bending, the individual unit cells of the crystal structure will deform. Such deformation will cause shifting and/or broadening of the diffracted line profile from a given reflection. Use of two measurements, one in the correct Bragg position and one slightly displaced, avoids the need for an unstressed comparison sample, and allows the individual principal stresses to be calculated [18]. The X-ray method is particularly useful for the determination of residual stress because it is applicable to the specimen as is, and does not involve pretreatment of the sample. Automated machines are available today that allow the rapid measurement of stress both on laboratory test specimens and on large finished pieces such as weldments in pipe lines, aircraft wings, and so on.

REFERENCES

1. Hanawalt, J. D., and Rinn, H. W. (1936). *Ind. Eng. Chem. Anal. Ed.* **8**, 244.
2. Hanawalt, J.D., Rinn, H.W., and Frevel, L. K. (1938). *Ind. Eng. Chem. Anal. Ed.* **10**, 457.
3. Guinier, A. (1939). *Ann. Phys. (Paris)* **12**, 161.
4. Bigelow, W. C., and Smith, J. V. (1965). *ASTM Spec. Tech. Publ.* **54**, 54.
5. Frevel, L. K. (1965). *Anal. Chem.* **37**, 471.
6. Johnson, G. G., and Vand, V. (1967). *Ind. Eng. Chem.* **59**, 19.
7. Frevel, L. K., Adams, C. E., and Ruhberg, L. R. (1976). *J. Appl. Cryst.* **9**, 300.
8. Edmonds, J. W. (1980). *J. Appl. Cryst.* **13**, 191.
9. Jenkins, R. (1977). *Adv. X-Ray Anal.* **20**, 125.
10. Jenkins, R., and Hubbard, C. R. (1978). *Adv. X-Ray Anal.* **22**, 133.
11. Jenkins, R., et al. (Feb. 1983). "Automated powder diffractometry, new dimensions in instrumentation and analytical software", *Norelco Reporter*, special issue.
12. Marquart, R. G., et al. (1979). *J. Appl. Cryst.* **12**, 629.
13. Jenkins, R., and Holomany, M. (1987). *Powder Diff.* **2**, 215.
14. Marquart, R. G. (1986). *Powder Diff.* **1**, 34.
15. Klug, H. P., and Alexander, L. E. (1974). *X-Ray Diffraction Procedures*, 2nd ed., Wiley: New York.

16. Calvert, L. D., Sirianni, A. F., Gainsford, G. J. and Hubbard, C. R. (1982). *Adv. X-ray Anal.* **26**, 105.
17. Fawcett, T. G., et al. (1985). *Adv. X-ray Anal.* **29**, 323.
18. Gisen, F., Glocker, R., and Osswald, E., (1936). *Z. Tech. Phys.* **17**, 145.

INSTRUMENTATION FOR X-RAY FLUORESCENCE SPECTROMETRY

4.1 HISTORICAL DEVELOPMENT OF X-RAY SPECTROMETRY

X-ray fluorescence spectrometry provides the means for the identification of an element by measurement of its characteristic X-ray emission wavelength or energy. The method allows the quantification of a given element by first measuring the emitted characteristic line intensity and then relating this intensity to elemental concentration. While the roots of the method go back to the early part of this century, it is only during the last 25 years or so that the technique has gained major significance as a routine means of elemental analysis.

The first use of the X-ray spectrometric method dates back to the classic work of Henry Moseley in 1912 [1]. In Moseley's original X-ray spectrometer, the source of primary radiation was a cold cathode tube in which the source of electrons was residual air in the tube itself, with the specimen for analysis forming the target of the tube. Radiation produced from the specimen then passed through a thin gold window onto an analyzing crystal, where it was diffracted to the detector. One of the major problems in the use of electrons for the excitation of characteristic X-radiation is that the process of conversion of electron energy into X-rays is inefficient: about 99% converted to heat energy. This means in turn that it may be difficult to analyze specimens which are volatile or tend to melt. Nevertheless, the technique seemed to hold some promise as an analytical tool, and an early paper on the use of X-ray spectroscopy for real chemical analysis appeared as long ago as 1922, in which Hadding [2] described the use of the technique for the analysis of minerals. In 1925, a practical solution to the problems associated with electron excitation was suggested by Coster and Nishina [3], who used primary X-ray photons for the excitation of secondary characteristic X-ray spectra. The use of X-rays, rather than electrons, to excite characteristic X-radiation avoids the problem of the heating of the specimen. It is possible to produce the primary X-ray photons inside a sealed X-ray tube under high vacuum and efficient cooling conditions, which means the specimen itself need not be subject to heat dissipation problems or the high vacuum requirements of the electron beam system. Use of X-rays rather than electrons represented the beginnings of the technique of X-ray fluorescence as we know it today.

The fluorescence method was first employed on a practical basis in 1928 by Glocker and Schreiber [4]. Unfortunately, data obtained at that time were rather poor because X-ray excitation is rather inefficient relative to electron excitation, and the detectors and crystals available at that time were rather primitive. Thus the fluorescence technique did not seem to hold much promise. In the event, widespread use of the technique had to wait until the mid 1940's, when it was rediscovered by Freidman and Birks [5]. The basis of their spectrometer was a diffractometer that had originally been designed for the orientation of quartz oscillator plates. A Geiger counter was used as a means of measuring the intensities of the diffracted characteristic lines, and quite reasonable sensitivity was obtained for a very large part of the atomic number range.

The first commercial X-ray spectrometer became available in the early 1950s, and although the earlier spectrometers operated only in air path conditions, they were able to provide qualitative and quantitative information on all elements above atomic number 22 (titanium). Later versions allowed use of helium or vacuum paths that extended the lower atomic number cutoff. Most modern spectrometers allow the determination of elements down to atomic number 9 (fluorine), and with special precautions even down to carbon (atomic number 6). Today, nearly all commercially available X-ray spectrometers use the fluorescence excitation method and employ a sealed X-ray tube as the primary excitation source. Some of the simpler systems may use a radioisotope source, because of considerations of cost and/or portability. While electron excitation is generally not used in "standalone" X-ray spectrometers, it is the basis of X-ray spectrometry carried out on electron column instruments. The ability to focus the primary electron beam allows analysis of extremely small areas down to a micron or so in diameter. This, in combination with imaging and electron diffraction, offers an extremely powerful method for the examination of small specimens, inclusions, grain boundary phenomena, etc. The instruments used for this type of work may be in the form of a specially designed electron microprobe analyzer [6] or simply an energy or wavelength dispersive attachment to a scanning electron microscope.

Among major developments in X-ray fluorescence instruments in the early 1960's were the use of lithium fluoride as a diffracting crystal and the use of the chromium and rhodium target X-ray tubes, which were especially useful for the excitation of longer wavelengths. Multichannel spectrometers also became available at this time, in which many spectrometers were grouped around the specimen, allowing the simultaneous measurement of many elements, albeit with some loss of flexibility. The computer controlled spectrometer became available in the mid 1960s. Probably the most significant development in recent years came in 1970 with the advent of the Si(Li) lithium drifted silicon detector (see, e.g., ref. 7). This detector gives a very high energy resolution and provides the means of separating the X-ray photons coming from an excited specimen without recourse to the use of a relatively inefficient analyzing crystal.

There are many types of X-ray fluorescence spectrometer available on the market today, but most of them fall roughly into two categories: wavelength dispersive instruments and energy dispersive instruments. The two approaches are illustrated in figure 4-1. In wavelength dispersive spectrometers the diffracting property of a single crystal is used to separate the polychromatic beam of radiation coming from the specimen. In energy dispersive spectrometers a Si(Li) detector is utilized to give a spectrum of voltage pulses that is directly proportional to the spectrum of X-ray photon energies entering the detector. An electronic voltage level sorter, called a multichannel analyzer, is then used to separate and collect these voltage pulses and record them in terms of their energies. The wavelength dispersive system was introduced commercially in the early 1950s, and probably around 15,000 such instruments have been supplied, roughly half of them in the U.S.A. Energy dispersive spectrometers became commercially available in the early 1970s, and today there are several thousands of these units in use.

4.2 EXCITATION OF X-RAYS

Several different types of source have been employed for the excitation of characteristic X-radiation; the more important of these are illustrated in figure 4-2. As has been mentioned previously, all of the early work in X-ray

a) Wavelength Dispersive

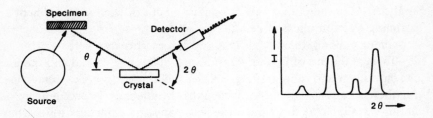

1) Single crystal of fixed 2d acts as a spectrum analyzer.

2) Scanning 2θ range allows the complete spectrum to be acquired.

3) Selection of single wavelength is achieved by selection of equivalent 2θ value.

b) Energy Dispersive

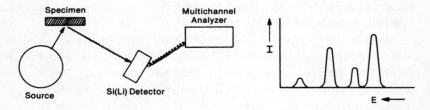

1) Proportional Si(Li) detector gives a distribution of voltage pulses proportional to the spectrum of X-ray photons

2) A multichannel analyzer is used to isolate the voltage pulses into discrete intervals. Consecutive output of the MCA intervals allows complete spectrum to be displayed.

3) Selection of a single energy interval is obtained by selection of appropriate voltage window (i.e., range of channels) on the MCA.

Figure 4-1. The wavelength dispersive spectrometer and the energy dispersive spectrometer.

spectrometry was done using electron excitation, and this technique is still used very successfully today in electron column applications. Not much use is made of electron excitation in classical X-ray fluorescence systems, due mainly to the inconvenience of having to work under high vacuum and to problems of heat dissipation. By far the most common source today is the X-ray photon source. This source is used in primary mode in the wavelength and primary energy dispersive systems, and in secondary fluorescer mode in secondary target energy dispersive spectrometers. The γ-source is typically a

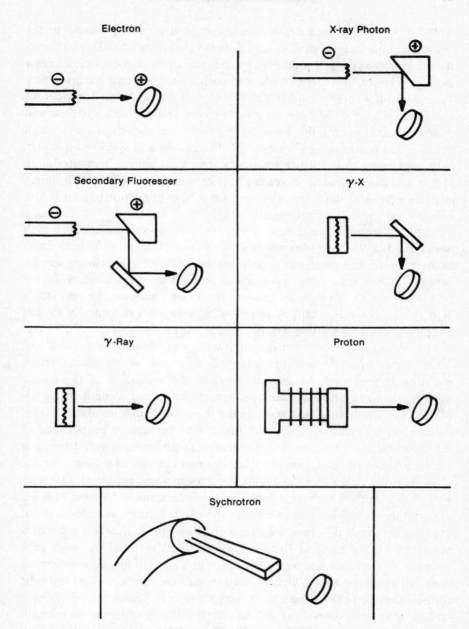

Figure 4-2. Sources for the excitation of characteristic X-radiation.

radioisotope that is used either directly, or in a mode equivalent to the secondary, fluorescer mode in energy dispersive spectrometry. This configuration is known as the γ-X source: γ-rays are used to excite X-radiation from a secondary target, and this secondary radiation used to excite the specimen. Typical γ-sources are ^{241}Am, ^{109}Cd, ^{153}Gd, ^{155}Eu and ^{145}Sm. The γ-source and the γ-X source are both used in low cost and portable systems, and though the total photon flux is small, they both offer the advantages of high stability, compactness, and low cost [8]. The proton and synchroton source both offer enormously higher sensitivity than the classical bremsstrahlung source, and each is finding increasing application in areas where high source intensity is an advantage. (See chapter 5 for further discussion of proton and synchrotron source spectrometers.)

Most conventional wavelength dispersive X-ray spectrometers use a high power (2–4 kW) X-ray bremsstrahlung source. Energy dispersive spectrometers use either a high power or low power (0.5–1.0 kW) primary source, depending on whether the spectrometer is used in the secondary or the primary mode. In all cases, the primary source unit consists of a very stable high voltage generator, capable of providing a potentia of typically 40–100 kV, plus a sealed X-ray tube. The X-ray tube has an anode of Cr, Rh, W, Ag, Au or Mo and delivers intense, continuous and characteristic radiation, which then impinges onto the analyzed specimen, where characteristic radiation is generated. In general, most of the excitation of the longer wavelength characteristic lines comes from the longer wavelength characteristic lines from the tube, and most of the short wavelength excitation, from the continuous radiation from the tube. Since the relative proportions of characteristic to continuous radiation from a target increase with decreasing atomic number of the anode material, optimum choice of a target for the excitation of a range of wavelengths can present some problems. One way around this problem is to use a dual target tube, and many different varieties of such tubes have been employed over the years. A recent manifestation of this is a dual anode tube in which the second (low atomic number) material is plated on top of the first (high atomic number) material. At high tube voltages the electrons penetrate beyond the thin layer of low atomic number material, and the output is highly biased toward the continuum. At low tube voltages the electrons dissipate their energy mainly in the low atomic number surface layer, with a resulting output biased in favor of longer wavelength radiation. Combinations of Sc and Mo, Cr and Ag, and Sc and W have all proven useful in this regard [9].

In order to excite a given characteristic line, the source must be run at a voltage V_0 well in excess of the critical excitation potential V_c of the element

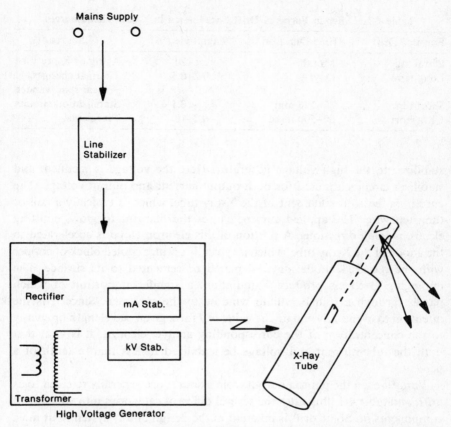

Figure 4-3. Basic components of the primary X-ray photon source using a sealed X-ray tube.

in question. The relationship between the measured intensity I of the characteristic line, the tube current i and the operating and critical excitation potentials is as follows:

$$I = Ki(V_0 - V_c)^{1.6} \qquad (4\text{-}1)$$

The product of i and V_0 represents the maximum output of the source in kilowatts. The optimum value for V_0/V_c is 3 to 5. This optimum value occurs because at very high operating potentials, the electrons striking the target in the X-ray tube penetrate so deep into the target that self-absorption of target radiation becomes significant. Figure 4-3 shows the basic components of a typical high power source. Power from the mains supply is fed via a line

Table 4-1. Different Forms of Drift Encountered in X-Ray Tube Sources

Form of Drift	Time Duration	Magnitude (%)	Source
Ultralong	Months	1–20	Aging of X-ray tube
Long term	Days	0.2–0.5	Thermal changes, focal spot wander
Short term	30–120 min	<0.1	Stabilization circuits
Ultrashort	50–500 msec	0.2–10	Transients

stabilizer to the high voltage generator. Here the voltage is rectified, and stabilizer circuits are used for both output current and output voltage. The current is fed to the filament of the X-ray tube, which is typically a coil of tungsten wire. The applied current causes the filament to glow, emitting electrons in all directions. A portion of this electron cloud is accelerated to the anode of the X-ray tube, which is typically a water cooled block of copper with the required anode material plated or cemented to its surface. The impinging electrons produce X-radiation, a significant portion of which passes through a thin beryllium window to the specimen. Since it is the intention to eventually equate the value of I for a given wavelength or energy to the concentration of the corresponding analyte element, it is vital that both the tube current and voltage be stabilized to less than a tenth of a percent.

Variations in the photon output from a source are generally referred to as *drift*, and table 4-1 illustrates the several different categories into which drift components fit. Some drift is inherent in the design of any system, but most of it is correctable. With such correction measures, significant systematic errors may accrue in the quantitative analysis. Most commercially available high voltage generators are designed to give very high stability over a short period of time (typically 30–120 minutes). The associated random error due to source instability is referred to as *short-term drift* and is typically of the order of a tenth of one percent or less. This is the limiting error in any quantitative procedure. There is also a *long-term drift* component that arises from a variety of sources, including thermal effects on components, focal spot wander in the tube, etc. The long-term drift is typically 2–5 times the short-term drift. *Ultrashort-term drift* arises where the stabilization circuits are unable to react quickly enough to short duration changes in the input power supply. Even a modern solid state stabilizer may have a reaction time of the order of a few hundred milliseconds. There is also an *ultralong-term drift* component that is really aging of the X-ray tube, due mainly to deposition of tungsten from the filament onto the inner surfaces of the tube.

While ultrashort-term drift can only be eliminated by addition of extra stabilization or use of an isolation transformer, the effects of both long and ultralong-term drift can be completely removed using the method of ratio counting. The basis of the ratio counting technique is to measure the counting rate on an instrument reference standard at frequent intervals, during the process of taking counts on specimens being analyzed. This instrument reference is typically a specimen selected from the suite of specimens used in the calibration of the spectrometer for the analytical problem in question. Ideally, it should contain all analyte elements, at concentration levels falling somewhere in the middle of the range being covered. By ratioing count rates from the analyte elements in the unknown specimen to those similarly obtained from the instrument reference, most of the effects of instrument and source long-term drift can be eliminated.

4.3 DETECTION OF X-RAYS

An X-ray detector is a transducer for converting X-ray photon energy into voltage pulses. Detectors work through a process of photoionization in which interaction between the entering X-ray photon and the active detector material produces a number of electrons. The current produced by these electrons is converted to a voltage pulse by a capacitor and resistor, such that one digital voltage pulse is produced for each entering X-ray photon.

In addition to being sensitive to the appropriate photon energies (i.e. being applicable to the needed range of wavelengths or energies), there are two other important properties that an ideal detector should possess. These properties are *proportionality* and *linearity*, as illustrated in figure 4-4. In the first part of the figure (a), an X-ray photon of energy E_1 enters the detector and a pulse of V_1 volts is produced. If V is proportional to E, the detector is said to be *proportional*. Proportionality is needed where the technique of pulse height selection is to be used. Pulse height selection is a means of electronically rejecting pulses of voltage levels other than those corresponding to the characteristic line being measured. This technique is a very powerful tool in reducing background levels and the influence of overlapping lines from elements other than the analyte [10].

The property of linearity is illustrated in the lower part of the figure (b). Here, a number of X-ray photons are entering the detector at a rate of I_1 photons per second, producing pulses at a rate R_1 pulses per second. If R is

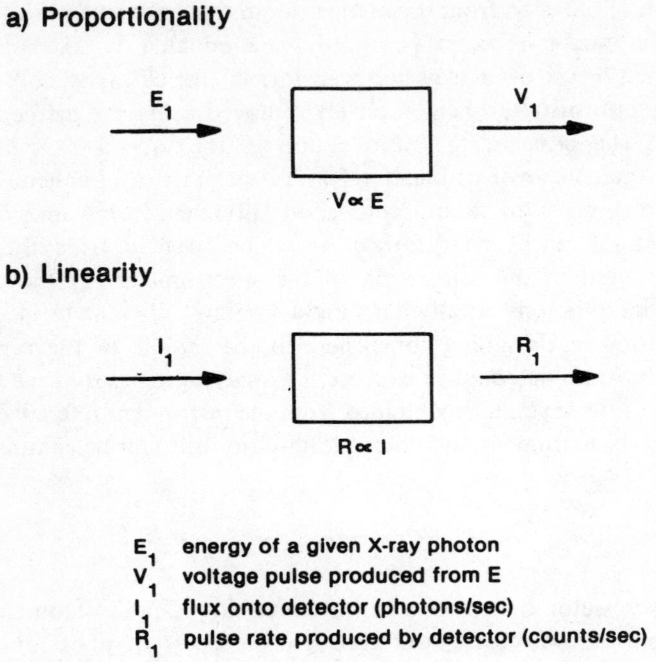

a) Proportionality

$E_1 \longrightarrow$ ▭ $\longrightarrow V_1$

$V \propto E$

b) Linearity

$I_1 \longrightarrow$ ▭ $\longrightarrow R_1$

$R \propto I$

E_1 energy of a given X-ray photon
V_1 voltage pulse produced from E
I_1 flux onto detector (photons/sec)
R_1 pulse rate produced by detector (counts/sec)

Figure 4-4. Proportionality and linearity in X-ray detectors.

proportional to *I*, the detector is said to be *linear*. Linearity is important where the various count rates produced by the detector are to be used as measures of the photon intensity for each measured line.

The properties of proportionality and linearity are, to a certain extent, controllable by the electronics associated with the actual detector. To this extent, while it is common practice to refer to the characteristics of detectors, the properties of the associated pulse processing chain should always be included.

A gas flow proportional counter, shown schematically in figure 4-5, consists of a cylindrical tube about 2 cm in diameter, carrying a thin (25–50 μm) wire along its radial axis. The tube is filled with a mixture of inert gas and quench gas—typically 90% argon and 10% methane (P-10). The cylindrical tube is grounded and around 1400–1800 volts is applied to the central wire. The wire is connected to a resistor R shunted by a capacitor C. An X-ray photon entering the detector produces a number *n* of ion pairs,

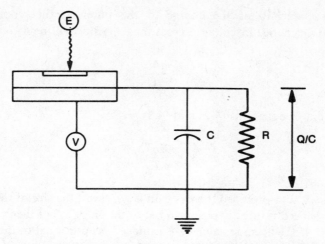

Figure 4-5. Schematic diagram of the gas flow proportional detector.

each composed of one electron and one positive argon ion. The first ionization potential for argon is about 16 eV, but competing processes during the conversion of photon energy to ionization cause the average energy required to produce an ion pair to be greater than this amount. The average energy to produce one ion pair divided by the first ionization potential is called the Fano factor F [11]. For argon, F is between $\frac{1}{2}$ and $\frac{1}{3}$, and the average energy e_i required to produce one primary ion pair is equal to 26.4 eV. Thus the number of ion pairs produced by a photon of energy E will equal

$$n = E/e_i \qquad (4\text{-}2)$$

Following ionization, the charges separate, with the electrons moving towards the (anode) wire and the argon ions to the grounded cylinder. As the electrons approach the high field region close to the anode wire, they are accelerated sufficiently to produce further ionization of argon atoms. Thus a much larger number N of electrons will actually reach the anode wire. This effect is called *gas gain*, or *gas multiplication*, and its magnitude is given by $M = N/n$. For gas flow proportional counters used in X-ray spectrometry M typically has a value of around 10^5. Thus the final charge Q appearing on the capacitor, for each incident photon, will be equal to $N \times q_e$, where q_e is the elementary charge, equal to 1.60×10^{-19} coulombs. Since capacitance (C) is defined as the ratio of the charge to the potential difference V, the corresponding change in voltage across the capacitor will be equal to

$$V = \frac{nMq_e}{C} \qquad (4.3)$$

Combination of equations (4-2) and (4.3) gives:

$$V = \frac{E}{e_i} \frac{q_e}{C} M \qquad (4-4)$$

It will be seen that provided the gas gain is constant, the size of the voltage pulse V produced is directly proportional to the energy E of the incident X-ray photon. If the average value of V is indeed proportional to the incident photon energy, the counter is said to be *proportional*.

In practice, not all photons arising from photon energy E will produce pulses of height exactly equal to V. There is a random process associated with the production of the voltage pulses, and the resolution of a counter is related to the variance in the average number of ion pairs produced per incident X-ray photon. The resolution is generally expressed in terms of the full width at half maximum of the pulse amplitude distribution. The theoretical resolution R_t of a flow counter can be derived from

$$R_t(\%) = \frac{2.36 \times 100}{\sqrt{n}} \qquad (4-5)$$

Combining equations (4-2) and (4-5) gives

$$R_t = \frac{38.3}{\sqrt{E}} \qquad (4-6)$$

While the gas flow proportional counter is ideal for the measurement of longer wavelengths, it is rather insensitive to wavelengths shorter than about 1.5 Å. For this shorter wavelength region, it is common to use the scintillation counter. The scintillation counter consists of two parts, the phosphor (scintillator) and the photomultiplier. The phosphor is typically a large single crystal of sodium iodide that has been doped with thallium. When X-ray photons fall onto the phosphor, blue light photons are produced, and their number is related to the energy of the incident X-ray photon. These blue light photons produce electrons by interaction with a photosurface in the photomultiplier, and the number of electrons is linearly increased by a series of secondary surfaces, called *dynodes*, in the photomultiplier. The current produced by the photomultiplier is then converted to a voltage pulse, as in the case of the gas flow proportional counter. Since the number of electrons is proportional to the energy of the incident X-ray photon, the scintillation counter is a *proportional counter*.

Because of inefficiencies in the X-ray blue-light electron conversion processes, the average energy to produce a single event with the scintillation counter is more than an order of magnitude greater than the equivalent process in the flow counter. For this reason, the resolution of the scintillation counter is much worse than that of the flow counter. The theoretical resolution R_t of the scintillation counter for photons of energy E is given by

$$R_t = \frac{128}{\sqrt{E}} \qquad (4\text{-}7)$$

The gas scintillation counter combines the principles of both gas and scintillation counters [12]. In this case, as in the case of the gas proportional counter, the incident X-ray photons first ionize the counter gas. The electrons are accelerated by a voltage across two wire mesh electrodes, where electron atom collisions cause the emission of ultraviolet radiation. This light is measured with a photomultiplier as in the case of the scintillation detector. Because there are less losses associated with these processes, in the gas scintillation counter than in the conventional scintillation counter, the gas scintillation counter gives about a threefold improvement in energy resolution. There are indication that the characteristics of the gas scintillation counter can be improved even further, at which time it may well find a use in conventional wavelength dispersive spectrometers.

The Si(Li) detector consists of a small cylinder (about 1 cm in diameter and 3 mm thick) of p-type silicon that has been compensated by lithium to increase its electrical resistivity [13]. A Schottky barrier contact on the front of the silicon disk produces a p–i–n type diode. In order to inhibit the mobility of the lithium ions and to reduce electronic noise, the diode and its preamplifier are cooled to the temperature of liquid nitrogen. While some progress has recently been reported on the use of miniature refrigerator compressors for this cooling process [14], the common practice remains to mount the detector assembly on a "cold finger" that is in turn kept cool with a reservoir of liquid nitrogen. By applying a reverse bias of around 1000

Table 4-2. Characteristics of Common X-Ray Detectors

Detector Type	Useful Range	Average Energy for One Ion Pair (eV)	Electrons/Photon Cu $K\alpha$		
			Initial	Gain	Final
Gas flow proportional	1.5–50	26.4	305	6×10^4	2×10^7
Scintillation	0.2–2	350	23	10^6	2×10^7
Si(Li)	0.5–8	3.8	2116	1	2×10^3

volts, most of the remaining charge carriers in the silicon are removed. Incident X-ray photons interact to produce a number n of electron–hole pairs, as given in equation (4-2), where, in this instance, e_i is equal to about 3.8 eV for cooled silicon. This charge is swept from the diode by the bias voltage to a charge sensitive preamplifier. A charge loop integrates the charge on a capacitor C to produce an output pulse, as in the case of the flow proportional counter, although in this case the factor M in equation (4-3) is equal to unity, since the Si(Li) detector does not have property equivalent to gas gain.

The resolution R of the Si(Li) detector is given by

$$R = [(\sigma_{\text{noise}})^2 + (2.35\{e_i FE\}^{1/2})^2]^{1/2} \qquad (4\text{-}8)$$

The Fano factor for the Si(Li) detector is around 0.12, and the noise contribution about 100 eV. Using a value of 3.8 for e_i, the calculated resolution from equation (4-8) for Mn Kα radiation ($E = 5{,}895$ eV) is about 160 eV.

Table 4-2 summarizes some of the characteristics of the three detectors commonly employed in X-ray spectrometry, and table 4-3 compares the resolution of these detectors with that of the crystal spectrometer. As was shown in equation (4-2), the number of (equivalent) ion pairs produced is directly proportional to the energy of the X-ray photon and inversely proportional to the average energy to produce one ion pair. In going from the Si(Li) detector to the flow counter to the scintillation counter, the average energy to produce an ion pair increases by roughly an order of magnitude at each step. Since the resolution of the detector is proportional to the square root of the number of electrons per photon, one would expect the resolution of the three detectors to differ by roughly the square root of 10, or about

Table 4-3. Resolution of Various Dispersion Devices at the Energy of Cu Kα (8.04 keV)

	Resolution (eV)	Resolution (%)
(a) Detector Alone		
Scintillation	3638	45.3
Gas Flow Proportional	1086	13.5
Si(Li)	160	2.0
(b) Crystal Spectrometer[a]		
LiF (200)	31	0.39
LiF (220) 1st order	22	0.27
LiF (220) 2nd order	12	0.15

[a] Based on a 10-mm-long primary collimator of 150 μm spacing and a 5-cm-long secondary collimator with 120-μm spacing.

three times. As shown in table 4-3(a), the data for Cu Kα show such a variation.

Another important practical point, shown in table 4-2, is the actual number of electrons produced, again in this example, for a photon of Cu Kα. The size of the voltage pulse produced by the cathode follower of the detector is proportional to the current produced, i.e. to the number of electrons reaching the collector. In a detector that has internal gain, this number of electrons will be the product of the initial number of electrons and the gain of the detector. In the case of the gas flow proportional counter, this is the gas gain, and in the scintillation counter it is the photomultiplier gain. It will be remembered that the Si(Li) detector has no internal gain. The Si(Li) detector has a high number of initial electrons per photon (therefore, a high resolution), but, relative to the other two detectors, has a small number of final electrons. This means that normal external amplification of voltage pulses from the Si(Li) detector cannot be used, because this would also amplify noise from the detector. For this reason, a cooled, charge-sensitive preamplifier is used in Si(Li) spectrometers rather than a simple linear electronic amplifier.

Table 4-3 shows the actual resolution of the detectors used alone, in comparison with a crystal spectrometer. As will be shown in the next section, the resolution of a crystal spectrometer is related to a number of factors, of which the *d*-spacing of the crystal and the order of the reflection are the most important. While the resolution of the Si(Li) detector is worse than the crystal spectrometers, in the middle of the usual analytical wavelength region it is sufficient for the separation of most lines. (Data here are for Cu Kα; as the energy of the analyte photon increases, the resolution difference between detectors alone and crystal spectrometers get less.) Note that the resolution of both gas flow and scintillation counters is almost never sufficient for use without a crystal spectrometer. (See section 6.3 for a more detailed discussion of the relative merits of wavelength and energy dispersive spectrometers for the separation of characteristic lines.)

4.4 WAVELENGTH DISPERSIVE SPECTROMETERS

The function of the spectrometer is to separate the polychromatic beam of radiation coming from the specimen in order that the intensities of each individual characteristic line can be measured. A spectrometer should provide sufficient resolution of lines to allow such data to be taken, at the same time providing a sufficiently large response above background to make the measurements statistically significant, especially at low analyte concentration levels. It is also necessary that the spectrometer allow measurements over the wavelength range to be covered. Thus, in the selection of a set of spectrometer operating variables, four factors are important: resolution, response, background level and range. For many reasons, optimum selection of some of these characteristics may be mutually exclusive; as an example, attempts to improve resolution invariably cause lowering of absolute peak intensities.

The wavelength dispersive spectrometer may be a single channel instrument in which a single crystal and a single detector are used for the measurement of a series of wavelengths sequentially, or a multichannel one in which many crystal detector sets are used to measure many elements simultaneously. In the typical wavelength dispersive spectrometer geometry, a single crystal of known interplanar spacing is used to disperse the collimated polychromatic beam of characteristic wavelengths coming from the sample, so that each wavelength will diffract at a discrete angle. As was shown in equation (1-8), there is simple relationship between the interplanar spacing of the crystal, the diffracted wavelength, and the diffraction angle. Since the maximum achievable angle on a typical wavelength dispersive spectrometer is around $\theta = 73°$, the maximum wavelength that can be diffracted by a crystal of spacing $2d$ is equal to about $1.9d$. In terms of the separating power of a crystal spectrometer this will be dependent upon two factors—the divergence allowed by the collimators (which to a first approximation determines the width of the diffracted lines), and the angular dispersion of the crystal [15]. The experimental breadth B_e of a line in a crystal spectrometer can be expressed as

$$B_e = (B_{cc}^2 + B_{cr}^2)^{1/2} \qquad (4\text{-}9)$$

where B_{cc} is the angular aperture of the primary collimator and B_{cr} the rocking curve of the diffracting structure. The wavelength resolution $\Delta\lambda/\lambda$ is a function of B_e and the angle θ at which the line is diffracted:

$$\frac{\Delta\lambda}{\lambda} = \frac{B_e}{\tan\theta} \qquad (4\text{-}10)$$

Typical values for B_e range from $2\theta = 0.1°$ at low angles to $0.5°$ at high angles, giving $\Delta\lambda/\lambda$ values in the range 0.002 to 0.02. In terms of energy, this corresponds to approximately 10–100 eV.

From the above, it will be clear that there is about one order of magnitude variation in the resolution of a wavelength dispersive spectrometer over the usual wavelength range, depending upon the selection of crystals and collimators. Since mechanical limitations prevent wide selectability of line shape through selection of collimator divergence, in practice the resolution of the spectrometer will mainly be a function of the angular dispersion of the analyzing crystal, albeit with some influence of the breadth of the diffracted line profile. The angular dispersion $d\theta/d\lambda$ of a crystal of spacing $2d$ is given by

$$\frac{d\theta}{d\lambda} = \frac{n}{2d\cos\theta} \qquad (4\text{-}11)$$

It will be seen from equation (4-11) that the angular dispersion will be high when the spacing d is small. This is unfortunate as far as the range of the spectrometer is concerned, because a small value of $2d$ means a small range of wavelengths coverable. Thus, as with the resolution and peak intensities, high dispersion can only be obtained at the expense of cutting down the wavelength range covered by a particular crystal. In order to circumvent this problem, it is likely that several analyzing crystals will be employed in the coverage of a number of analyte elements. Many different analyzing crystals

**Table 4-4. Analyzing Crystals Used in
Wavelength Dispersive X-Ray Spectrometry**[a]

Crystal	Planes	2d (Å)	Atomic Number Range	
			K lines	L lines
LiF	(220)	2.848	>Ti(22)	>La(57)
LiF	(200)	4.028	>K(19)	>Cd(48)
PE	(002)	8.742	Al(13)–K(19)	—
TAP	(001)	26.4	F(9)–Na(11)	—
LSM	—	50–120	Be(4)–F(9)	—

[a] For a comprehensive list of analyzing crystals see Bertin, E. P., (1975), *Principles and Practice of X-ray Spectrometric Analysis*, 2nd ed., Plenum: New York, Appendix 10.

are available, each having its own special characteristics (see e.g. ref. 16), but three or four crystals will generally suffice for most applications. Table 4-4 shows a short list of the more commonly used crystals. While the maximum wavelength covered by traditional spectrometer designs is about 20 Å, recent developments now allow the extension of the wavelength range significantly beyond this value. For a further discussion of this refer to section 5.3.

The actual spectrometer itself consists of the X-ray tube, a specimen holder support, a primary collimator, an analyzing crystal and a tandem detector, all mounted on a goniometer. The geometric arrangement of these components is shown in figure 4-6. A portion of the characteristic "fluorescence" radiation from the specimen is passed via a collimator or slit onto the surface of an analyzing crystal, where individual wavelengths are diffracted to the detector in accordance with Bragg's law. A goniometer is used to maintain the required θ to 2θ relationship between crystal and detector. Typically six or so different analyzing crystals and two different collimators are provided in this type of spectrometer, giving the operator a wide range of choice of dispersion conditions. In general, the smaller the spacing d of the crystal, the better the separation of the lines, but the smaller the wavelength range that can be covered. A tandem detector system is typically employed, comprising a gas flow counter and a scintillation counter, each with its own collimator, and this is used to convert the diffracted characteristic photons into voltage pulses, which are integrated and displayed as a measure of the characteristic line intensity. The gas flow counter is ideal for the measurement of the longer wavelengths, and the scintillation counter is best for the short wavelengths. It was recognized some time ago that, since the bremsstrahlung arising from scatter from the primary

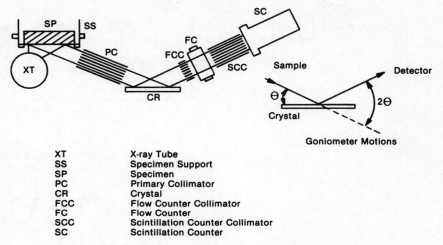

XT	X-ray Tube
SS	Specimen Support
SP	Specimen
PC	Primary Collimator
CR	Crystal
FCC	Flow Counter Collimator
FC	Flow Counter
SCC	Scintillation Counter Collimator
SC	Scintillation Counter

Figure 4-6. The wavelength dispersive spectrometer.

X-ray source is scattered twice (once by the sample and once by the analyzing crystal), it is plane polarized [17]. Fluorescence radiation from the sample is scattered only once and is therefore not polarized. Advantage can be taken of this by arranging the geometry of the source, sample and analyzing crystal to reduce the amount of scattered radiation from the sample entering the detector [18]. More recently, it has been shown that the use of a curved crystal further increases the efficiency of the background reduction, leading to detection limits in the sub-ppm range [19].

The output from a wavelength dispersive spectrometer may be either analog or digital, and as illustrated in figure 4-7, both digital and analog counting equipment is generally available to the user. Pulses from the detector are amplified and passed via the pulse height selector on to one of two circuits. For qualitative work an analog output is traditionally used and in this instance a ratemeter is used to integrate the pulses over short time intervals, typically of the order of a second or so. The output from the rate meter is fed to an x t recorder that scans at a speed that is synchronously coupled with the goniometer scan speed. The recorder thus displays an intensity time diagram, which becomes an intensity 2θ diagram. Tables are then used to interpret the wavelengths. For quantitative work it is more convenient to employ digital counting, and a timer–scaler combination is provided that will allow pulses to be integrated over a period of several tens

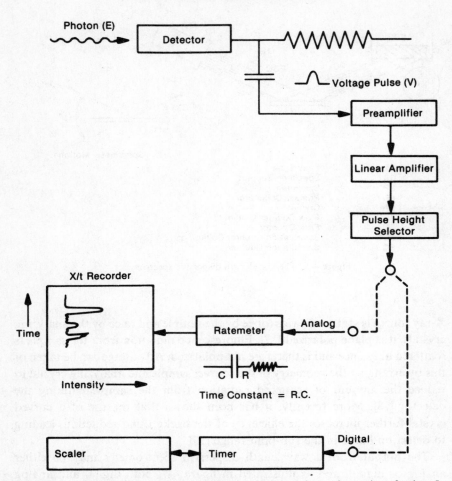

Figure 4-7. Analog and digital instrumentation for the integration and counting of pulses. In fixed time counting, a present time t generates a stop pulse to the scaler. In fixed counting, a present count on the scaler generates a stop pulse to the timer.

of seconds and then displayed as count or count rate. In more modern spectrometers a timer scaler may also take the place of the rate meter, using a process known as *step scanning*, illustrated in figure 4-8. In this instance the contents of the scaler are displayed on the *x* axis of the x t recorder as a voltage level. The scaler timer is then reset and started to count for a selected time interval. At the end of this time the timer sends a stop pulse to the scaler, which now holds a number of counts equal to the product of the counting

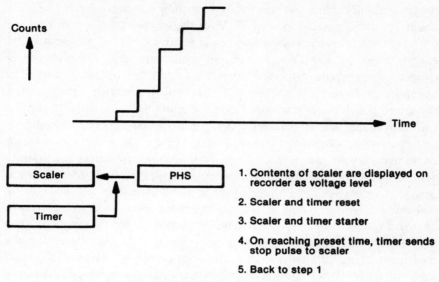

Figure 4-8. Step scanning to display spectra.

rate and the count time. The contents of the scaler are then displayed as before, the goniometer stepped to its next position and the whole cycle repeated. Generally the process is completely controlled by a microprocessor or a minicomputer.

4.5 ENERGY DISPERSIVE SPECTROMETERS

Unlike the wavelength dispersive spectrometer, the energy dispersive spectrometer consists of only two basic units: the excitation source and the spectrometer detection system, since the detector itself acts as the dispersion agent. The spectrometer detector is typically a Si(Li) detector which, as was shown in section 4.3, is a proportional detector of high intrinsic resolution. A multichannel analyzer is used to collect, integrate and display the resolved pulses.

While the same properties are sought in the energy dispersive system as in the wavelength dispersive system, the means of selecting these optimum conditions are very different. Since the resolution of the energy dispersive system is equated directly to the resolution of the detector, this feature is of paramount importance. As was shown in section 4.3, this value for Si(Li) is about 160 eV, compared to 10–100 eV for the wavelength dispersive system (see chapter 6 for further discussion of the relative merits of wavelength and energy dispersive systems). The output from an energy dispersive spectrometer is generally displayed on a CRT, and the operator is able to dynamically display the contents of the various channels as an energy spectrum. Provision is generally made for zooming in on portions of the spectrum of special interest, overlaying spectra, subtracting the background, and so on, in an interactive manner. As in the case of the modern wavelength dispersive systems, nearly all energy dispersive spectrometers incorporate some form of minicomputer for spectral stripping, peak identification, quantitative analysis, and a host of other useful functions.

Even though the Si(Li) detector is the most common detector used in energy dispersive X-ray spectrometry, it is certainly not the only one. The higher absorbing power of germanium makes it an alternative for the measurement of high energy spectra, and both cadmium telluride (CdTe) and mercuric iodide (HgI_2) show some promise as detectors capable of operating satisfactorily at room temperature. There are, however, many practical problems in the manufacture of these devices. As an example, in the preparation of CdTe, a Br–methanol etchant is used, and the side product $CdBr_2$ has been found to poison the surface, causing high leakage current with increase in background noise. This problem has recently been solved, and much lower backgrounds are now being reported [20]. The major problem with HgI_2 is in growing crystals of suitable size. The resolution of the detector is, however, promising, making the rather tedious research worth while. As an example, a HgI_2 based energy dispersive spectrometer, in which both detector and FET were cooled using a Peltier cooler, had been used in a scanning electron microscope, and a resolution of 225 eV (FWHM) obtained for Mn Kα (5.9 keV) and 195 eV for Mg Kα (1.25 keV) [21].

All of the earlier energy dispersive spectrometers were operated in what is called the *primary* mode and consisted simply of the excitation source (typically a closely coupled low power, end window X-ray tube) and the detection system. In principle, this primary excitation system offered the possibility of a relatively inexpensive instrument, with two significant advantages over the wavelength dispersive system: firstly, the ability to collect and display the total emission spectrum from the sample at the same time, giving great speed in the acquisition and display of data, and secondly, mechanical simplicity in that there is almost no need at all for moving parts. In practice, however, there is a limit to the maximum count rate that the spectrometer can handle, and this led, in the mid-1970s, to the development of the *secondary* mode of operation (see e.g. ref. 22). In the secondary mode, a carefully selected pure element standard is interposed between primary source and specimen, along with absorption filters where appropriate, such that a selectable energy range of secondary photons is incident upon the sample. This allows selective excitation of certain portions of the energy range, thus increasing the fraction of useful to unwanted photons entering the detector. While this configuration does not completely eliminate the count rate and resolution limitations of the primary system, it certainly does reduce them.

As has been previously mentioned, within the two major categories of X-ray spectrometers specified, there is a wide diversity of instruments available. The major differences generally lie in the type of source used for excitation, the number of elements that are measurable at one time, the speed at which they collect data, and finally the price range. All of the instruments are, in principle at least, capable of measuring all elements in the periodic classification from $Z = 9$ (F) upwards. Most can be fitted with multisample handling facilities, and all can be automated by the use of minicomputers. All are capable of precisions of the order of a few tenths of one percent, and all have sensitivities down to the low ppm level. As far as the analyst is concerned, they differ only in their speed, cost and number of elements measurable at the same time.

REFERENCES

1. Moseley, H. G. J. (1912). *Phil. Mag.* **26**, 1024; (1913), **27**, 703.
2. Hadding, A. (1922). *Z. Anorg. Allgem. Chem.* **122**, 195.
3. Coster, D., and Nishina, J. (1925). *Chem. News* **130**, 149.
4. Glocker, R., and Schreiber, H. (1928). *Ann. Physik* **85**, 1085.
5. Birks, L. S. (1976). *History of X-Ray Spectrochemical Analysis*, American Chemical Society Centennial Volume, ACS, Washington.
6. Birks, L. S. (1963). *Electron Probe Microanalysis*, Chemical Analysis Series, Vol. XVII, Interscience: New York.
7. Jenkins, R., Gould, R. W. and Gedcke, D. A. (1981). *Quantitative X-ray Spectrometry*, Dekker: New York, section 4.3.
8. Rhodes, J. R. (1971). "Design and application of X-ray emission analysers using radioisotopes." *In Energy Dispersive X-Ray Analysis*, ASTM Special Technical Publication 485, ASTM: Philadelphia.
9. Kikkert, J. N., and Hendry, G. (1983). *Adv. X-Ray Anal.* **27**, 423.

10. Jenkins, R., and de Vries, J. L. (1970). *Practical X-ray Spectrometry*, 2nd ed., MacMillan: London, chapter 4.

11. Fano, U. (1946). *Phys. Rev.* **70**, 44.

12. Ku, W. H. M., and Novick, R. (1981). *AIP Conf. Proc.* **75**, 78.

13. Gedcke, D. A. (1972). *X-Ray Spectrom.* **1**, 129.

14. Kevex Corporation Cryoelectric Detector (1987). Kevex Corporation, Foster City, California.

15. Jenkins, R. (1974). *An Introduction to X-Ray Spectrometry*, Wiley/Heyden: London, chapter 4.

16. Bertin, E. P. (1975). *Principles and Practice of X-Ray Spectrometric Analysis*, 2nd ed., Plenum: New York, Appendix 10.

17. Barkla, C. G. (1906). *Proc. Roy. Soc.* **A77**, 247.

18. Wobrauschek, P., and Aiginger, H. (1985). *Adv. X-Ray Anal.* **28**, 69.

19. Gilfrich, J. V., Shelton, E. F., Quadri, S. B., Kirkland, J. P., and Nagel, D. J. (1983). *Anal. Chem.* **55**, 187.

20. Roth, M., and Burger, A. (1986). *IEEE Trans. Nucl. Sci.* NS-33, 407.

21. Iwanczyk, J. S., Dabrowski, A. J., Huth, G. C., Bradley, J. G., Conley, J. M., and Albee, A. L. (1986). *IEEE Trans. Nucl. Sci.* 355.

22. L. S. Birks and H. K. Herglotz (Eds.) (1978). *X-Ray Spectrometry*, Dekker: New York, chapter 2.

CHAPTER

5

RECENT TRENDS IN X-RAY FLUORESCENCE INSTRUMENTATION

5.1 THE ROLE OF X-RAY FLUORESCENCE IN INDUSTRY AND RESEARCH

Over the past thirty years or so, the X-ray fluorescence method has become one of the most valuable methods for the qualitative and qualitative analysis of materials. Many methods of instrumental elemental analysis are available today, and among the factors that will generally be taken into consideration in the selection of one of these methods are accuracy, range of application, speed, cost, sensitivity and reliability. While it is certainly true is that no one technique can ever be expected to offer all of the features that a given analyst might desire, the X-ray method has good overall performance characteristics. In particular, the speed, accuracy and versatility of X-ray fluorescence are the most important features among the many that have made it the method of choice in over 15,000 laboratories all over the world. Both the simultaneous wavelength dispersive spectrometer and the energy dispersive spectrometers lend themselves admirably to the qualitative and quantitative analysis of solid materials and solutions. Because the characteristic X-ray spectra are so simple, the actual process of allocating atomic numbers to the emission lines is also simple, and the chance of making a gross error is rather small. The relationship between characteristic line intensity and elemental composition is also now well understood, and if intensities can be obtained that are free from instrumental artifacts, excellent quantitative data can be obtained. Today, conventional X-ray fluorescence spectrometers allow the rapid quantitation of all elements in the periodic table from fluorine (atomic number 9) upwards. Recent advances in wavelength dispersive spectrometers have extended this element range down to carbon (atomic number 6). Over most of the measurable range, accuracies of a few tenths of one percent are possible, with detection limits down to the low ppm level.

5.2   SCOPE OF THE X-RAY FLUORESCENCE METHOD

As was indicated in the previous section, the basis of the X-ray fluorescence technique lies in the relationship between the wavelength (or energy) of the X-ray photons emitted by the sample element and the atomic number Z, as given in Moseley's law, equation (1-1). Measurement of the wavelength with the crystal dispersive spectrometer, or of the energy with the energy dispersive spectrometer, allows the identification of the elements present. The intensities of the characteristic lines can then be used to calculate the concentrations of the elements. Most commercially available X-ray spec-trometers have a range from about 0.4 to 20 Å (40 to 0.6 keV), and this range will allow measurement of the K series from fluorine ($Z = 9$) to lutecium ($Z = 71$), and the L series from manganese ($Z = 25$) to uranium ($Z = 92$). Other line series can occur from the M and N levels, but these have little use in analytical X-ray spectrometry. As was discussed in section 1.3, because of the competing Auger process, the number of vacancies resulting in the production of characteristic X-ray photons is less than the total number of vacancies created in the excitation process. The ratio of the useful to total vacancies is the fluorescent yield, which takes a value from around unity for the higher atomic numbers to less than 0.01 for the low atomic number elements such as sodium, magnesium and aluminum. This is an important factor in determining the absolute number of counts that an element will give, under a certain set of experimental conditions. It is mainly for this reason that the sensitivity of the X-ray spectrometric technique is rather poor for the very low atomic number elements.

There is a wide variety of instrumentation available today for the application of X-ray fluorescence techniques, but for the purpose of this discussion on trends in instrumentation development it is useful to break X-ray spectrometers down into three main categories:

1. Wavelength dispersive spectrometers
   a. Scanning (sequential)
   b. Multichannel (simultaneous)
2. Energy dispersive spectrometers

    a. Primary excitation
    b. Secondary target
    c. Isotope excitation
3. Special spectrometers
    a. Total reflection (TRXRF)
    b. Synchrotron source (SSXRF)
    c. Proton induced (PIXE)

Wavelength dispersive spectrometers are by far the most commonly employed, and these systems employ diffraction by a single crystal to separate characteristic wavelengths emitted by the sample. Energy dispersive spectrometers use the proportional characteristics of a photon detector, typically lithium drifted silicon, to separate the characteristic photons according to their energies. Since there is a simple relationship between wavelength and energy, these techniques each provide the same basic type of information, and the characteristics of the two methods differ mainly in their relative sensitivities and the way in which data are collected and presented. Generally speaking the wavelength dispersive system is one to two orders of magnitude more sensitive that the energy dispersive system. Against this, however, the energy dispersive spectrometer measures all elements within its range at the same time, whereas the wavelength dispersive system identifies only those elements for which it is programmed. To this extent, the energy dispersive system is more useful in recognizing unexpected elements. Although, in principle, almost any high energy particle can be used to excite characteristic radiation from a specimen, a sealed X-ray tube offers a reasonable compromise between efficiency, stability and cost, and almost all commercially available X-ray spectrometers use that excitation source. The exception to this might be in the analysis of very thin specimens, where a proton source offers significant advantages [1]. Both wavelength and energy dispersive spectrometers typically employ a primary X-ray photon source operating at 0.5 to 3 kW. As discussed in section 1.5, the specimen scatters the bremsstrahlung from the source, leading to significant background levels that are among the major limitations in the determination of low concentration levels.

    The wavelength and energy dispersive spectrometers of the type discussed thus far are typical of what might be found in a modern analytical laboratory. The analysis time required varies from about 10 seconds to 3 minutes per element. The minimum sample size required is of the order of a few milligrams, although typical sample sizes are probably several grams.

The accuracies obtainable are excellent, and in favorable cases standard deviations of the order of a few tenths of one percent are possible. This is because the matrix effects in X-ray spectrometry are well understood and fairly easy to overcome. The sensitivity is fair, and determinations down to the low parts per million level are possible for most elements. All elements above atomic number 9 (fluorine) are measurable by this technique.

The third category considered here consists of special X-ray spectrometers, which, while they may not be generally available to the general user community, do have important roles to play in special areas of application. Included within this category are total reflection spectrometers (TRXRF), sychrotron source spectrometers (SSXRF), and proton induced X-ray emission spectrometers (PIXE). Two things that each of these three special systems have in common are a very high sensitivity and ability to work with extremely low concentrations and/or small specimens.

The TRXRF system makes use of the fact that at very low glancing angles, primary X-ray photons are almost completely absorbed within thin specimens, and the high background that would generally occur due to scatter from the sample support is absent.

The recent development of high intensity synchrotron radiation beams has led to interest in their application for X-ray fluorescence analysis. The high intensity of these beams allows use of very narrow band path monochromators, giving, in turn, a high degree of selective excitation. This selectivity overcomes one of the major disadvantages of the classical EDS approach and allows excellent detection limits to be obtained.

The proton induced X-ray emission system differs from conventional energy dispersive spectroscopy in that a proton source is used in place of the photon source. The proton source is typically a Van de Graaff generator or a cyclotron. Protons in the energy range of about 2–3 MeV are typically employed for this type of work. The proton source offers several advantages over the photon source. In addition to being much more intense source, the proton excitation system generates much lower backgrounds. In addition, the cross section for characteristic X-ray production is quite large, and good excitation efficiency is possible.

5.3   THE DETERMINATION OF LOW ATOMIC NUMBER ELEMENTS

Classically, large single crystals have been used as diffracting structures in the wavelength dispersive spectrometer. The three dimensional lattice of atoms is fabricated and oriented so that Bragg planes form the interatomic spacing $2d$

for the wavelength in question. Modern spectrometers employ a number of different crystals, each with its specific $2d$ spacing, in order to cover the full wavelength range under conditions of optimum dispersion. The selection of crystals for the longer (>8 Å) wavelength region is difficult, however, mainly because there are not many crystals available for work in this region. The most commonly employed crystal is probably TAP, thallium acid phthalate ($2d = 26.3$ Å), and this allows measurement of the K lines of elements below atomic number 13 (Al). While the lower atomic number elements, including magnesium, sodium, fluorine and oxygen, can be measured with this crystal, unfortunately its reflectivity is rather weak, resulting in rather poor sensitivity. In addition to this, the crystal tends to deteriorate over a period of many months, causing the reflection to drop by as much as 50%.

Several alternatives to single crystals as diffracting structures have been sought over the years, including the use of complex organic materials with large $2d$ spacings, gratings, and specular reflectors; metal disulfides; and organic intercalation complexes such as graphite, molybdenum disulfide, mica, and clays. Moderate success for long wavelength measurements was achieved by the use of soap films [2] having spacings in the range 80 to 120 Å. These layered structures are composed of planes of heavy metal cations separated by chains of organic acids. The basis of their usefulness as diffracting structures is the periodic electron density contrast between the heavy metal sites and the lower density organic material. Lead octadeconate (LOD), with a $2d$ spacing around 100 Å, is one such film that has been used for carbon and oxygen analysis. Unfortunately, this diffracting medium has neither adequate angular dispesion nor adequate reflectivity for the elements fluorine, sodium and higher atomic numbers.

In addition to crystals and multilayer films, a third alternative has recently become available, namely the *layered synthetic microstructure* (LSM). LSM's are constructed by applying successive layers of atoms or molecules on a suitably smooth substrate. In this manner, both the $2d$ spacing and the composition of each layer are selected for optimum diffraction characteristics. Figure 5-1 illustrates the three principal types of diffracting structures, fabricated of natural single crystals, multilayer films and layered synthetic microstructures.

The first successful attempts to make LSM's were made in 1940 by DuMond and Youtz [3] and later by Dinklage and Frericks [4]. However, while these early workers were partially able to overcome problems of deposition of the layers, the resulting structures were unstable to interdiffusion of the layers. During the development of X-ray normal incidence mirrors, Spiller was able to advance the technology significantly [5] by

a) **Natural Single Crystal**

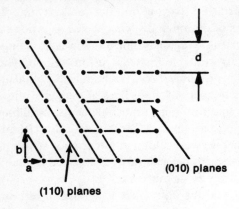

(010) planes

(110) planes

1. **Regular three dimensional array of atoms.**

   Crystal is oriented and cut in a given crystallographic direction (hkl).

2. **Diffraction occurs due to the planes of high atomic density.**

3. **Diffraction planes correspond to interplanar spacing for selected (hkl), in the example, the (010) set of planes.**

b) **Multilayer Film**

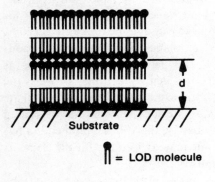

Substrate

∏ = LOD molecule

1. **Long chain molecules (e.g. lead octadecanate LOD) with heavy atom sites deposited in a specific orientation on to substrate.**

2. **Diffraction occurs due to high electron density contrast between heavy atoms and rest of molecule.**

3. **Diffraction planes correspond to twice the length of the chain of the molecule.**

c) **Layered Synthetic Microstructure**

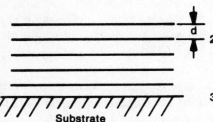

Substrate

1. **Successive layers of low and high atomic number elements deposited onto substrate.**

2. **Diffraction occurs due to high electron density contrast between low and high atomic number deposits.**

3. **Diffraction planes correspond to the average distance between successive low and high atomic number layers.**

Figure 5-1. Three principal types of diffracting structures.

working with exceptionally smooth substrates and by careful control of layer thickness by in situ monitoring of X-ray reflectivity. Parallel developments by Barbee and associates made use of sputtering technology in the synthesis of large multilayer mirrors. This work resulted in better control of composition and layer thickness. Both Barbee [6] and Henke [7] have presented performance data for these LSM's that by far exceed those obtainable with TAP and LOD. A special feature of LSM's is that, to a certain extent, they can be designed and fabricated to give optimum performance for special applications [8]. Henke [9] has described procedures for the detailed characterization of multilayer analyzers that can be effectively coupled to their design, optimization and application.

Figure 5-2 shows peak diffraction coefficients of LSM's compared with LOD and conventional diffracting structures. The solid curves exhibit the peak intensity ratios using three commercially available LSM's [10] referred to as PX1, PX2 and PX3. The PX1 LSM has a $2d$ spacing of 50 Å, and the PX2 a $2d$ spacing of 120 Å. These data show factors of about 4.5 improvement in peak intensities over TAP, for the range of elements measured. For very low atomic numbers, including boron, carbon, nitrogen and oxygen, the PX2 structure shows peak intensity improvements of about a factor of 6 over to LOD. Table 5-1 shows data obtained for the elements beryllium ($Z = 4$), boron ($Z = 5$) and carbon ($Z = 6$) using LSM's [11] and indicates that detection limits of the order of tenths of a percent are achievable.

5.4   TOTAL REFLECTION X-RAY FLUORESCENCE—TRXRF

One of the major problems that inhibits the obtaining of good detection limits in small samples is the high background due to scattering from the sample substrate support material (see section 1.6). The suggestion to overcome this problem by using total reflection of the primary beam was made as long ago as 1971 [12], but the absence of suitable instrumentation prevented real progress being made until the late 1970s [13, 14]. Mainly due to the work of Schwenke and his co-workers, good sample preparation and presentation procedures are now available, making TRXRF a valuable technique for trace analysis [15]. The more recent availability of commercial instruments (e.g. ref. 16) is likely to further enhance the application of this method to a wide variety of problems.

The TRXRF method is essentially an energy dispersive technique in which

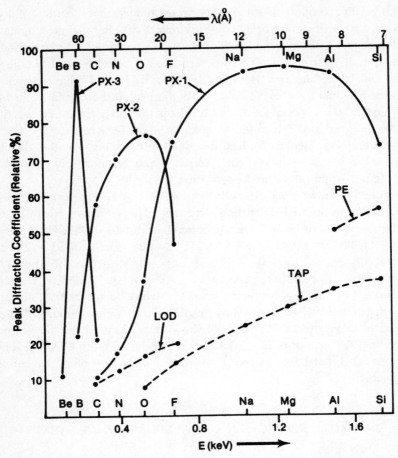

Figure 5-2. Peak diffraction coefficients for penta-erythritol (PE), lead octa-decanoate (LOD) and thallium acid phthalate (TAP) structures (dashed lines), and for layered synthetic microstructures PX-1, PX-2 and PX-3 (solid lines).

the Si(Li) detector is placed close to (about 5 mm), and directly above, the sample. Primary radiation enters the sample at a glancing angle of a few seconds of arc. The sample itself is typically presented as a thin film on the optically flat surface of a quartz plate. In the instrument described by Michaelis et al. [15] a series of reflectors is employed to aid in the reduction of background; see figure 5-3. Here, a beam of radiation from a sealed X-ray tube passes through a fixed aperture onto a pair of reflectors that are placed

Table 5.1. Data Obtained with Low Atomic Number Elements Using Layered Synthetic Microstructures

Element	Sensitivity M (counts/sec %)	R_b (counts/sec)	Matrix	LLD[a] (%)
Be (4)	0.11	16	Be foil	1.1
B (5)	5	45	Borosilicate glass	0.4
C (6)	22	200	Coal	0.2

[a]Statistical lower limits of detection calculated for a total counting time t of 200 seconds, i.e. $t_b = t/2 = 100$ seconds: $LLD = (3/M)\sqrt{R_b/t_b}$.

very close to each other. Scattered radiation passes through the first aperture to impinge on the sample at a very low glancing angle. Because the primary radiation enters the sample at an angle barely less than the critical angle for total reflection, this radiation barely penetrates the substrate media; thus scattering and fluorescence from the substrate are minimal. Because the background is so low, picogram amounts can be measured or concentrations in the range of a few tenths of a ppb can be obtained in aqueous solutions without recourse to preconcentration [17].

One area in which the TRXRF technique has found great application is in the analysis of natural waters. The concentration levels of, for example, transition metals in rain, river and sea waters are normally too low to allow estimation by standard X-ray fluorescence techniques, unless preconcentration is employed (see section 7.6). Using TRXRF, concentration levels down to less than 10 µg/L are achievable.

While the TRXRF method is most applicable to homogeneous liquid samples, success has also been achieved in the application of the method to solids including particulates, sediments, air dusts and minerals. In these instances the sample is first digested in concentrated nitric acid and then diluted to a calibrated volume with ultrapure water, after the addition of an internal standard. Where undissolved material is still present, this may be dispersed using an ultrasonic bath before the specimen is taken. In addition to the advantages of ease of specimen preparation and the ability to handle milligram quantities of material, the TRXRF method is also relatively simple to apply quantitatively. Because the specimen is only a few microns thick, one generally does not observe the rather complicated matrix effects usually encountered with thick samples. Thus the only standard required is the one needed to establish the sensitivity of the spectrometer, in counts/sec per

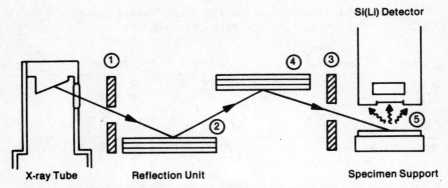

Figure 5-3. Schematic representation of the total reflection X-ray spectrometer: 1, first aperture; 2, first reflection unit; 3, second aperture; 4, second reflection unit; 5, specimen.

percent, for the element(s) in question. This standard is generally added to the sample to be analyzed during the specimen preparation procedure.

5.5 SYNCHROTRON SOURCE X-RAY FLUORESCENCE—SSXRF

The availability of intense, linearly polarized synchrotron radiation beams [18] has prompted workers in the fields of X-ray fluorescence (e.g. ref. 19) and X-ray diffraction (e.g. ref. 20) to explore what the source has to offer over more conventional excitation media. In the synchrotron, electrons with kinetic energies of the order of several billion electron volts (typically 3 GeV at this time) orbit in a high vacuum tube between the poles of a strong (about 10^4 gauss) magnet. A vertical field accelerates the electrons horizontally, causing the emission of synchrotron radiation. Thus synchrotron source radiation can be considered as magnetic bremsstrahlung, in contrast to normal electronic bremsstralung produced when electrons are decelerated by the electrons of an atom. In the case of both fluorescence and diffraction it has been found that because the primary source of radiation is so intense, it is possible to use a high degree of monochromatization between source and specimen, giving a source that is wavelength (and therefore energy) tunable, as well as being highly monochromatic. There are several different excitation modes that can be used using SSXRF, including direct excitation with continuum, excitation with absorber modified continuum, excitation with source crystal monochromatized continuum, excitation with source radiation scattered by a mirror, and reflection and transmission modes. Figure 5-4 illustrates the more important of these source configurations. These various schemes have been studied to give optimum signal to noise ratio for various types of specimen [21, 22].

Additional advantages accrue because synchrotron radiation is highly polarized. In the case of EDXRF, the background due to coherent and incoherent scattering can be greatly reduced by placing the detector at 90° to the path of the incident beam and in the plane of polarization. The technique is so sensitive that it has been possible to make studies of concentration profiles at the liquid–gas interface of sulfonated and manganese neutralized polystyrene dissolved in DMSO [23]. A disadvantage of the SSXRF technique is that the source intensity decreases with time, but this can be overcome by bracketing analytical measurements between source standards and/or by continuously monitoring the primary beam. In addition to this, problems can also arise from the occurrence of diffraction peaks from highly ordered specimens. This has proven particularly troublesome in the analysis of small quantities [10–100 parts per 10^9 (ppb) weight in 20 µm spots] of geological specimens [24]. Giauque et al. described experiments [25] using the Stanford Synchrotron Radiation Laboratory to establish what minimum detectable limits (MDL's) could be obtained under optimum excitation conditions. Using thin film standards on stretched tetrafluoropolyethylene mounts, it was found that for counting times of the order of a few hundred seconds, MDL's of the order of 20 ppb could be obtained on a range of elements including Ca, Ti, V, Mn, Fe, Cu, Zn, Rb, Ge, Sn and Pb. The optimum excitation energy was found to be about one and a quarter times the absorption edge energy of the element in question. By working with a fixed excitation energy of 18 keV, the average MDL was found to be about 100 ppb. Synchrotron source radiation EDS is also readily applicable to very small specimen sizes; as an example, detection limits of less than 1 ppm have been obtained on a 2 µL droplet of air dried whole blood [25]. Other workers have reported detection limits in the range of 0.003–0.3 µg/cm² for a range of elements on thin film filters [18]. In combination with the TRXRF method, using a lapped silicon support, detection limits down to 0.5 ppb have been reported [26].

The intensity of the synchrotron beam is 4 to 5 orders of magnitude greater than that of the conventional bremsstrahlung source, sealed X-ray tubes. This, in combination with its energy tunability and polarization in the plane of the synchrotron ring, allows very rapid bulk analyses to be obtained on small areas. This is especially attractive for the analysis of archeometric

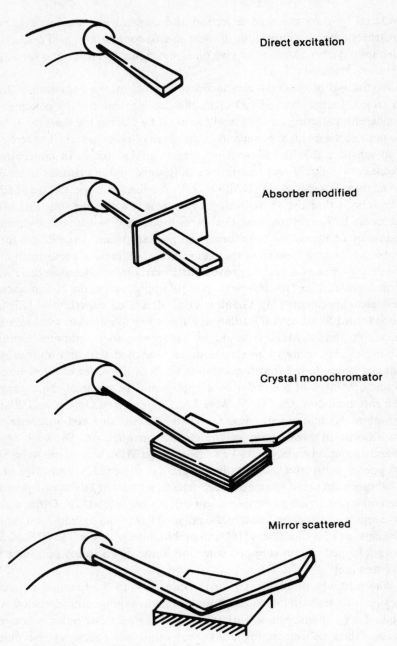

Figure 5-4. Source configurations used in synchrotron source excitation.

samples, for example, ancient ceramics [22]. Because the synchrotron beam has such a high intensity and small divergence, it is possible to use it as a high resolution (about 10 μm) microprobe. Absolute limits of detection around 10^{-14} g have been reported using such an arrangement [27]. Synchrotron source X-ray fluorescence has also been used in combination with TRXRF. Very high signal/background ratios have been reported employing this arrangement for the analysis of small quantities of aqueous solutions dried on the reflector, with detection limits of <1 ppb or 1 pg [21].

5.6 PROTON EXCITED X-RAY FLUORESCENCE—PIXE

While the use of protons as a potential source for the excitation of characteristic X-rays has been recognized since the early 1960s, it is only over the past several years that the technique has come into its own [28]. The PIXE method uses a beam of fast ions (protons) of primary energies in the range 1–4 MeV. In addition to the ion accelerator, the system contains an energy defining magnetic deflection field, a magnetic or electrostatic lens along the excitation beam pipe, a high vacuum target chamber for the specimen(s) and an energy dispersive detector analyzer. The great advantage of the PIXE method over other sources is that it generates only a small amount of background and is thus applicable to very low concentration levels and the analysis of very small samples. As an example, PIXE has been used for monitoring the iron, copper, zinc, bromine and lead contents of aerosol particulates [29]. The great sensitivity allowed integration times as short as 30–60 minutes. In another rather unusual application, PIXE has been used for the analysis of a large number of trace elements in wine samples [30].

The use of conventional X-ray fluorescence and PIXE have been compared with special reference to applications in art and archeology. Together, they seem to offer the museum scientist and archaeologist excellent complementary tools for nondestructive testing [31]. Comparison of the PIXE method has been made with many other spectroscopic techniques for the analysis of twenty-two elements in ancient pottery [32].

The high sensitivity of PIXE has also been of great use in the field of forensic science. In a recent example [33], external beam PIXE was used to verify the presence of lead in the finger bone of a murder victim. The deceased, who had been buried for several years, was known to have suffered a bullet wound in his right hand several years before his death. Analysis of the second right proximal phalanx, using a 1.5 MeV proton beam, showed a unique distribution of lead in an area of metal fragments indicated by X-radiographs. These data confirmed that the lead had come from a gunshot wound.

PIXE has also found valuable application in medicine [34]. Applications include in vitro analysis of trace elements in human body fluids and normal pathological tissues, and in vivo analysis of iodine in the thyroid, lead in the skeleton and cadmium in the kidney. Trace elements in blood serum of patients with liver cancer have also been studied by PIXE. The copper/zinc ratio was found to be significantly higher than in normal patients [35]. The Serum copper/zinc ratio is potentially useful in the diagnosis and prognosis of liver cancer.

REFERENCES

1. Klockenkamper, R., et al. (1987). *Fresenius Z. Anal. Chem.* **326**, 105.

2. Henke, B. L. (1965). *Adv. X-ray Anal.* **8**, 269.

3. DuMond, J., and Youtz, J. P. (1940). *J. Appl. Phys.* **11**, 357.

4. Dinklage, J., and Frericks, R. (1963). *J. Appl. Phys.* **34**, 1633.

5. Spiller, E. (1973). *Optik* **39**, 181.

6. Barbee, T. W. (1981). *In Proc AIP Symposium on Low Energy X-ray Diagnostics*, Monterey, California, pp. 131–145.

7. Henke, B. L. *Ibid*, pp. 146–155.

8. Barbee, T. W. (1985). *Superlattices Microstruct.* **1**, 311.

9. Henke, B. J., Vejio, Y. J., Tackaberry, R. E., and Yamada, H. T. (1985). *Proc. SPIE—Int. Soc. Opt. Eng.*, p. 583.

10. Nicolosi, J. A., Jenkins, R., Groven, J. P., and Merlo, D. (1985). *Proc. SPIE* **563**, 378.

11. Nicolosi, J. A., Groven, J. P., and Merlo, D. (1987). *Adv. X-ray Anal.* **30**, 183.

12. Yoneda, Y., and Horiuchi, T. (1971). *Rev. Sci. Instrum.* **42**, 1069.

13. Knoth, J., and Schwenke, H. (1978). *Fresenius Z. Anal. Chem.* **301**, 200.

14. Schwenke, H., and Knoth, J. (1982). *Nucl. Instrum. and Methods* **193**, 239.

15. Michaelis, W., Knoth, J., Prange, A., and Schwenke, H. (1984). *Adv. X-Ray Anal.* **28**, 75.

16. Rich. Seiffert & Co., Röntgenwerk, D-2070 Ahrensburg, West Germany.

17. Aiginger, H., and Wobrauschek, P. (1985). *Adv. X-ray Anal.* **28**, 1.

18. Sparks, C. J., Jr. (1980). *In* H. Winnick and S. Doniach (Eds.), *Synchrotron Radiation Research*, Plenum: New York, p. 459.

19. Gilfrich, J. V., et al. (1983). *Anal. Chem.* **55**, 187.

20. Parrish, W., Hart, M., and Huang, T. C. (1986). *J. Appl. Cryst.* **19**, 92.

21. Iida, A. (1985). *Nippon Kessho Gakkaishi* **27**, 61.

22. Harbottle, G., Gordon, A. M., and Jones, K. W. (1986). *Nucl. Instrum. Methods Res. Sec. B.* **B14**(1), 116.

23. Bloch, J. M., et al. (1985). Brookhaven Natl. Lab. Rep., BNL-51847, p. 36.

24. Sutton, S. R., Rivers, M. L., and Smith, J. V. (1986). *Anal. Chem.* **58**, 2187.

25. Giauque, R. D., Jaklevic, J. M., and Thompson, A. C. (1986). *Anal. Chem.* **58**, 940–944.

26. Iida, A., and Goshi, Y. (1984). *Adv. X-Ray Anal.* **28**, 61.

27. Petersen, W., Ketelsen, P., Knoechel, A., and Pausch, R. (1986). *Nucl. Instrum. Meth. Phys. Sect. A* **A246**(1–3), 731.

28. Garten, R. P. H. (1984). *Trends in Anal. Chem.* **3**(6), 152.

29. Hietel, B., Schulz, F., and Wittmaack, K. (1986). *Nucl. Instrum. Meth. Phys. Res. Sect. B* **B15**(1–6), 608.

30. Houdayer, A., Hinrichsen, P. F., Martin, J. P., and Belhadfa, A. (1987). *Canad. Spectrosc.* **32**, 7.

31. Malmqvist, K. G. (1986). *Nucl. Instrum. Meth. Phys. Res. Sect. B* **B14**, 86.

32. Bird, J. R., Duerden, P., Clayton, E., Wilson, D. J., and Fink, D. (1986). *Ibid.* **B15**, 86.

33. Fischbeck, H. H., Ryan, S. R., and Snow, C. C. (1986). *J. Forensic Sci.* **31**, 79.

34. Cesareo, M. (1982). *X-Ray Fluorescence in Medicine*, Field Education Italia: Rome.

35. Lin, X., et al. (1985). *Zhonhna Zhongliu Zazhi* **7**, 411.

COMPARISON OF WAVELENGTH AND ENERGY DISPERSIVE SPECTROMETERS

6.1 INTRODUCTION

As has been discussed in earlier chapters, there are many different types of X-ray fluorescence instruments available, based on several different sources and dispersion methods. The most commonly employed instruments are based either on the energy dispersive method or on the wavelength dispersive method. Further subdivisions of the energy dispersive instruments include primary or secondary source excitation mode; of the wavelength dispersive instruments, single channel (sequential) or multichannel (simultaneous). All these instruments may or may not include microprocessor control, and may or may not include a data processing computer. While each of these configurations has clear advantages over its competitors, each also has disadvantages—other than the obvious ones of cost and inflexibility. Most instrument manufacturers continually strive to develop instruments that offer a good price/performance ratio and that minimize some of the inherent limitations of a given procedure. This chapter presents some of the advantages and disadvantages of energy and wavelength dispersive spectrometers.

Wavelength dispersive spectrometers were introduced in the early 1950s, and around 15,000 or so such instruments have been supplied commercially, roughly half of these in the U.S.A. Energy dispersive spectrometers became commercially available in the early 1970s, and today there are of the order of 4,000 units in use. When the first energy dispersive instruments were marketed, the wavelength dispersive method was already an established technique. Also, at this time, the digital minicomputer was being routinely employed in analytical instrumentation both for instrument control and for data acquisition and processing. It is was not surprising, therefore, that nearly all of the early energy dispersive systems incorporated state of the art minicomputers as an integral part of the system. At the same time, the use of graphical display monitors were also becoming commonplace, and these too were incorporated into these systems from the very early days. Alongside of its rather conservatively designed wavelength dispersive counterpart, the energy dispersive system presented a modern state of the art image, and many purchasers of those early machines were happy to overlook the count rate and resolution restrictions from which they suffered, in order to take advantage of the rapid data acquisition and display capabilities.

In the design of any spectrometer system the system designer is confronted with a series of choices and compromises if the end product is to meet the end users' requirements, at an acceptable price. This situation is especially important where the spectrometer is to be used for both qualitative and quantitative analysis. High speed of analysis combined with good sensitivity can be obtained for quantitative applications at the expense of flexibility, the multichannel wavelength dispersive spectrometer being a good example of such a compromise. Great flexibility can be obtained in qualitative analysis, on the other hand, at the expense of sensitivity and some accuracy, as typified by many energy dispersive systems.

For most laboratories there are three basic choices in the selection of an X-ray spectrometer:

a. For high specimen throughput quantitative analysis where speed is of the essence, and where high initial cost can be justified, simultaneous wavelength dispersive instruments are probably the best.
b. For more flexibility, where speed is important but not critical and where moderately high initial cost can be justified, sequential wavelength dispersive instruments are probably the best.
c. Where initial cost is a major consideration, or where something can be given up in detection limits or accuracy, or where qualitative or semiquantitative analysis is important, energy dispersive instruments are probably the best.

The two major categories of X-ray spectrometers differ mainly in the type of source used for excitation, the number of elements that they are able to measure at one time, the speed at which they collect data and their price range. All of the instruments are, in principle at least, capable of measuring all elements in the periodic classification from $Z = 9$ (F) and upwards, and most modern wavelength dispersive spectrometers can do some useful measurements down to $Z = 6$ (C). All can be fitted with multisample handling facilities, and all can be automated by use of minicomputers. All are capable of precisions of the order of a few tenths of one percent, and all have sensitivities down to the low ppm level. Single channel wavelength dispersive spectrometers are typically employed for both routine and nonroutine analysis of a wide range of products, including ferrous and nonferrous alloys, oils, slags and sinters, ores and minerals, thin films, and so on. These systems are very flexible but, relative to multichannel spectrometers, are somewhat slow. The multichannel wavelength dispersive instruments are used almost exclusively for routine, high throughput analyses where the great need is for fast accurate analysis, but where flexibility is of no importance. Energy dispersive spectrometers have the great advantage of being able to display information on all elements at the same time. In resolution they fall short of wavelength dispersive spectrometers, but the ability to reveal elements absent as well as elements present makes energy dispersive spectrometers ideal for general trouble shooting. They have been particularly effective in scrap alloy sorting, in forensic science and in the provision of elemental data to supplement X-ray powder diffraction data.

In recent years, new wavelength and energy dispersive spectrometers have evolved that offer good price–performance characteristics and each of which reflects the latest technology in its own discipline. An example is the development in the early 1980's of secondary target energy dispersive system. This system offers good price–performance characteristics within the medium price range. The response of the wavelength dispersive spectrometer developers has been to produce instruments in which price has been significantly reduced by sacrificing some of the great speed and sensitivity of the traditional sequential and simultaneous wavelength sytems, but that can still outperform energy dispersive type systems in many areas (see e.g. ref. 1).

6.2   THE MEASURABLE ATOMIC NUMBER RANGE

One of the problems with any X-ray spectrometer system is that the absolute sensitivity (i.e. the measured number of counts per second per percent) decreases dramatically as the lower atomic number region is approached (see section 8.5). The three main reasons for this are that the fluorescence yield decreases with decreasing atomic number, that the absolute number of useful long wavelength X-ray photons from a bremsstrahlung source decreases with increasing wavelength, and that absorption effects generally become more severe with increasing wavelength of the analyte line. The first two of these problems are inherent in the X-ray excitation process and in the basic design of conventional X-ray tubes. The third, however, depends very much on the instrument design, and in particular upon the characteristics of the detector. The detector that is used in long wavelength spectrometers is typically a gas flow proportional counter, in which an extremely thin, high transmission window is employed. The detector typically employed in energy dispersive systems is the Si(Li) diode, which has a electrical contact layer on the front surface—typically a 0.02 μm thick layer of gold, followed by a 0.1-μm-thick dead layer of silicon. The absorption problems caused by these two layers become most significant for low energy X-ray photons, which have a high probability of being absorbed in the dead layer. Probably the biggest cause of absorption loss in the Si(Li) detector is the thin beryllium window that is part of the liquid nitrogen cryostat. A combination of these facts causes a loss in the sensitivity of a typical energy dispersive system of almost an order of magnitude for the K lines on going from sulfur ($Z = 16$) to sodium ($Z = 11$). The equivalent factor with a gas flow counter is about two.

Until very recently, the lowest atomic number usefully detectable with a typical energy dispersive spectrometer has been $Z = 12$ (magnesium). New developments in ultrathin windows for the Si(Li) detector [2] now allow measurements down to $Z = 8$ (oxygen). Figure 6-1 shows data published by the KEVEX Corporation* comparing old (vertical bars) and new detector technology for the measurement of low atomic number elements in a geological specimen. In comparison, the lower atomic number limit for the conventional wavelength dispersive system is $Z = 9$ (fluorine), though by use of special crystals this can be extended down to $Z = 4$ (beryllium); see section 5.3.

6.3   THE RESOLUTION OF LINES

A compromise that must always be made in the design and setup of a spectrometer is that between intensity and resolution—resolution being defined as the ability of the spectrometer to separate lines. In a flat crystal wavelength dispersive system this resolution is dependent upon the angular dispersion of the analyzing crystal and the divergence allowed by the

---

*KEVEX Corporation, Foster City, California.

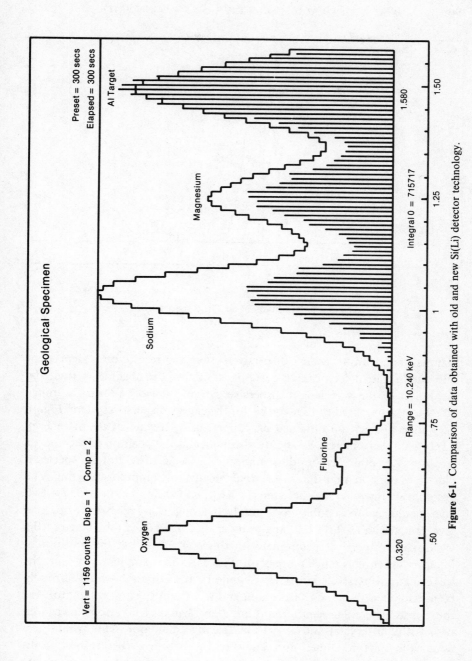

Figure 6-1. Comparison of data obtained with old and new Si(Li) detector technology.

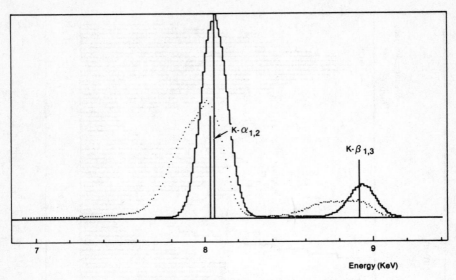

Figure 6-2. Resolution of the Si(Li) spectrometer.

collimators [3]. In an energy dispersive system the resolution is dependent only upon the detector and detector amplifier. In absolute terms, the resolution of the wavelength dispersive system typically lies in the range 10–100 eV, compared to 150–200 eV for the energy dispersive system. Figure 6-2 shows a spectrum obtained on copper, using a state of the art energy dispersive spectrometer. The figure also indicates the position of the copper Kα and Kβ doublets on the same energy scale. The dotted spectrum illustrates the potential shift of the doublets due to Compton scattering. The absolute energy difference between Cu $K\alpha_1$ and Cu $K\alpha_2$ is 21 eV. These two lines could be reasonably well resolved using a wavelength dispersive spectrometer and a LiF(220) analyzing crystal. An added advantage of the wavelength dispersive spectrometer in this context is that the resolution–intensity selection is much more controllable. As an example, when using the wavelength dispersive spectrometer, while Fe Kα radiation would normally be measured with a LiF(200) crystal and a 150 μm spacing collimator, for those cases where line separation is not a problem it could also be measured with LiF(200) crystal and a 450 μm spacing collimator. This would give about a factor three times more intensity, but with worsened resolution. In the case of secondary fluorescer systems some local modifications to the measured spectrum can be made by use of filters either in the primary beam or in the secondary fluoresced beam, and this does offer some flexibility.

6.4 MEASUREMENT OF LOW CONCENTRATIONS

X-ray fluorescence methods are reasonably sensitive, with detection limits for most elements in the low ppm range, (see also section 9.5). The lower limit of detection is generally defined as that concentration equivalent to a certain number of standard deviations of the background count rate. This means that three major factors will affect the detection limit for a given element: first, the sensitivity of the spectrometer for that element in terms of the counting rate per unit concentration of the analyte element; second, the background (blank) counting rate; and third, the available time for counting peak and background photons. In comparing energy and wavelength dispersive systems, the absolute sensitivity of the wavelength dispersive system is almost always higher than the equivalent value for an energy dispersive system—perhaps by one to three orders of magnitude. This is because modern wavelength dispersive systems are able to handle count rates up to 1,000,000 counts/sec, compared to about 40,000 counts/sec (for the total output of the selected excitation range) for most energy dispersive systems. The ability of the wavelength dispersive sytem to work at high counting rates allows use of a high loading at the primary source.

A fundamental difference between the wavelength and energy dispersive systems is that in the energy dispersive case, *all* of the detected radiation falls onto the detector at the same time. Figure 6-3 compares the cases of the energy and wavelength dispersive spectrometers measuring two elements i and j in a specimen, giving intensities I_i and I_j. In this example, the concentration of i is much greater than that of j, so that I_i is much greater than I_j. The total radiation flux falling onto the energy dispersive detector is $I_i + I_j$, whereas in the case of the wavelength dispersive system, I_i and I_j are measured independently at two different settings of the analyzing crystal. Since the detector is limited in the total radiation flux it can handle, the detection limit of j in i is limited by the total radiation intensity at the detector, and therefore the concentration, of i.

The influence of the background is rather complicated because so many variables come into play. One of the major advantages of the secondary target energy dispersive system over the conventional system is that backgrounds are dramatically lower in the secondary excitation mode, because much of the background in a fluoresced X-ray spectrum comes from scattered primary continuum. The measured backgrounds for the wavelength and energy dispersive systems are very similar in the low atomic number regions of the X-ray spectrum, and for the secondary target energy dispersive system they are lower by up to an order of magnitude in the higher

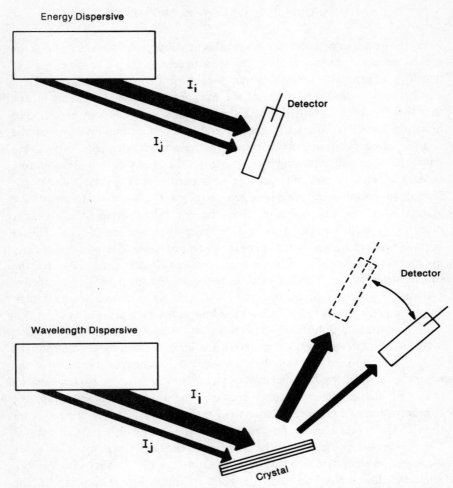

Figure 6-3. Count rate limitations with the energy dispersive spectrometer.

energy (midrange atomic number) region. All of these factors taken together lead to detection limits for the secondary target energy dispersive system that are typically a factor of 3–8 times worse than the wavelength dispersive spectrometer. The actual detection limit measurable also depends upon the characteristics of the specimen itself, including absorption, scattering power etc. Table 6-1 shows detection limits reported for trace and major elements in whole rock as reported for primary [4] and secondary [5] target energy

**Table 6-1. Detection Limits for Major and
Minor Elements in Whole Rock[a]**

Element	Detection Limit (reported as oxides)		
	PEEDS [ref. 4]	STEDS [ref. 5]	WDS [ref. 6]
	Major Elements[b]		
Na	0.96	0.81	0.16
Mg	0.33	0.13	0.08
Al	0.19	0.08	0.032
Si	0.21	0.023	0.050
P	0.05	0.023	0.016
K	0.04	0.022	0.008
Ca	0.025	0.034	0.004
Ti	0.03	0.008	0.006
Mn	0.015	0.002	0.014
	Minor Elements[c]		
Rb	5.6	3.0	0.6
Sr	3.5	2.8	0.4
Y	3.5	3.8	0.4
Zr	4.0	2.8	1.1
Nb	4.4	2.8	1.3

[a] PEEDS: primary excitation energy dispersive spectrometry; STEDS: secondary target energy dispersive spectrometry; WDS: wavelength dispersive spectrometry.
[b] Fused samples; detection limits in percent.
[c] Pressed powders; detection limits in ppm.

dispersive and wavelength dispersive spectrometers [6]. The trace elements data were obtained directly on the pressed rock powder, and the major element data were taken after fusion of the rock with lithium tetraborate. It will be seen that, on average, the secondary target system is better than the primary system by about a factor of four, and the wavelength dispersive data are better by a further factor of four.

6.5 QUALITATIVE ANALYSIS

One area in which energy dispersive systems have generally outperformed the wavelength dispersive system is in qualitative analysis. In wavelength dispersive systems, qualitative analysis has traditionally been performed by scanning the goniometer synchronously at a fixed angular speed, and then using a ratemeter circuit to integrate the digital output from the detector, displaying the output on an x/t recorder. All scanning spectrometers are slow in this sequential angular intensity data collection mode, not only because the data are taken sequentially, but also because in order to cover the full range of elements a series of scans must be made with different conditions. In addition to this, scanning at a fixed speed is somewhat inefficient because in the crystal dispersive system, atomic number varies as a function of one over the square root of the angle. In effect this means that in the low atomic number region much scanning time is wasted in scanning angular space that contains no characteristic line data. Although some special designs have brought about some reduction in data acquisition times [1], this remains a major limitation of the wavelength dispersive spectrometer. As far as data interpretation is concerned, although angle wavelength atomic number tables are readily available, interpretation is still tedious and time consuming. However, the count data output from modern wavelength dispersive systems is nearly always digital, and this means that state of the art spectral processing software can be used to speed up and automate the whole data interpretation process. Good qualitative software packages are available (see section 8.4), and a complete qualitative analysis can now be performed in about 45 minutes, while retaining all of the inherent resolution and sensitivity advantages of the wavelength dispersive system.

The output from an energy dispersive instrument is collected in a simultaneous fashion, either over the complete spectral range, or in selected portions, depending upon the complexity of the specimen and concentration levels involved. The operator is able to dynamically display the contents of the various channels as an energy spectrum, as they are acquired. Provision is generally made to allow interactively zooming in on portions of the spectrum of special interest, to overlay spectra, to subtract background, and so on. Nearly all energy dispersive spectrometers incorporate some form of mini-computer or microprocessor, which is available for spectral stripping, peak identification, quantitative analysis, and a host of other useful functions.

6.6 GEOMETRIC CONSTRAINTS OF WAVELENGTH AND ENERGY DISPERSIVE SPECTROMETERS

The modern wavelength dispersive system has tended to be somewhat inflexible in the area of sample handling. The geometric constraints of the conventional system generally stem from the need for close coupling of the distance from sample to X-ray tube, and the need to use an airlock of some kind to bring the sample into the working vacuum, usually by means of a multiple position sample carousel. The sample to be analyzed is typically placed inside a cup of fixed external dimensions that is in turn placed in the carousel. This presentation system places constraints not only on the maximum dimensions of the sample cup, but also on the size and shape of samples that can be placed in the cup itself.

Primary energy dispersive systems do not require the same high degree of focussing, and to that extent are more easily applicable to any sample shape or size, provided that the specimen will fit into the radiation protected chamber. In some instances the spectrometer can even be brought to the object to be analyzed. Because of this flexibility, the analysis of odd shaped specimens has been almost exclusively within the purview of the energy dispersive system.

In the case of secondary target systems, while the geometric constraints are still less severe than the wavelength system, they are much more critical than in the case of primary systems. This is due not just to the additional mechanical movements in the secondary target system, but also to the limitations imposed by the extremely close coupling of X-ray tube to secondary target. This is probably the reason that most energy dispersive spectrometer manufacturers generally offer a primary system for bulk sample analysis, and retain the secondary target system—generally equipped with a multiple specimen loader—for the analysis of samples that have been constrained to the internal dimensions of a standard sample cup during the specimen preparation procedure.

REFERENCES

1. Jenkins, R., Hammell, B., Cruz, A., and Nicolosi, J. A. (1985). *Norelco Reporter* **32**, 1.

2. Kevex Corporation (spring 1987). *New Product Bulletin*, Kevex: Foster City, California.

3. Jenkins, R. 1976. *An Introduction to X-Ray Spectrometry*, Wiley: London, Chapter 4, "Instrumentation."

4. Kevex Corporation (1987). *Kevex 0700 Laboratory X-Ray Fluorescence Spectrometer Brochure*, p. 5.

5. Kevex Corporation (1987). XRS Applications Report. *Geological Applications*, p. 2.1.6.

6. Jenkins, R., Nicolosi, J. A., Croke, J. F., and Merlo, D. (1985). *Norelco Reporter* **32**, 16.

CHAPTER
7

SAMPLE PREPARATION AND PRESENTATION

7.1 FORM OF THE SAMPLE FOR X-RAY ANALYSIS

There are many forms of sample suitable for X-ray fluorescence analysis, and the form of the sample as received will generally determine the method of pretreatment. It is convenient to refer to the material received for analysis as the *sample*, and that which is actually analyzed in the spectrometer as the *specimen*. While the direct analysis of certain materials is certainly possible, more often than not some pretreatment is required to convert the sample to the specimen. This step is referred to as specimen preparation.

In general, the analyst would prefer to analyze the sample directly, since if it is taken as received, any problems arising from sample contamination that might occur during pretreatment are avoided. In practice, however, there are two major constraints that may prevent this ideal circumstance from being achieved: sample size and sample heterogeneity. Problems of sample size are frequently severe in the case of bulk materials such as metals, large pieces of rock, etc. Problems of sample heterogeneity will generally occur under these circumstances as well, and in the analysis of powdered materials sample heterogeneity must almost always be considered.

Table 7-1 lists some typical sample forms and indicates possible methods of specimen preparation. Four types of sample form are listed—bulk solids, powders, liquids and gases. The sample as received may be either homogeneous or heterogeneous; in the latter case, it may be necessary to render the sample homogeneous before an analysis can be made. Heterogeneous bulk solids are generally the most difficult kind of sample to handle, and it may be necessary to dissolve or chemically react the material in some way to give a homogeneous preparation. Heterogeneous powders are either ground to a

103

Table 7-1. Forms and Treatments of Samples for Analysis

Form	Treatment
Bulk Solids	
Homogeneous	Grind to give flat surface
Heterogeneous	Dissolve or react to give solution or homogeneous melt
Powders	
Homogeneous	Grind and press into a pellet
Heterogeneous	Grind and fuse with borax
Liquids	
Homogeneous (concentrated)	Analyze directly or dilute
Homogeneous (dilute)	Preconcentration
Heterogeneous	Filter to remove solids
Gases	
Airborne dusts	Aspirate through a filter to remove solids

fine particle size and then pelletized, or fused with a glass forming material such as borax. Solid material in liquids or gases must be filtered out and the filter analyzed as a solid. Where analyte concentrations in liquids or solutions are too high or too low, dilution or preconcentration techniques may be employed to bring the analyte concentration within an acceptable range. These various specimen preparation techniques will be discussed in detail in succeeding sections of this chapter.

Most laboratory type spectrometers place constraints on the size and shape of the analyzed specimen. In general, the aperture into which the specimen must fit is in the form of a cylinder, typically 25–48 mm in diameter and 10–30 mm in height. Although the specimen that is placed in the spectrometer may be rather large, because of the limited penetration depth of characteristic X-ray photons, the actual mass of specimen analyzed is quite small. This is shown in figure 7-1. In the illustration, the specimen is represented as a disk of thickness T, density ρ and diameter $2r$. As was discussed in section 1.4, an estimation can be made of the approximate path length x travelled by characteristic X-ray photons, by assuming a certain fraction of absorbed radiation [see equation (1-5)]. It was also shown that the

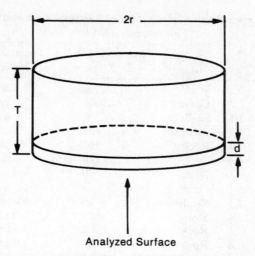

Figure 7-1. Mass of specimen actually contributing to the measured fluorescence intensity: T = speciment thickness, 2r = specimen diameter, d = penetration depth.

penetration depth d is related to the path length and the spectrometer takeoff angle ψ_2, as given in equation (1-6).

The mass of the sample analyzed, m_a, is given by

$$m_a = \pi r^2 d\rho \qquad (7\text{-}1)$$

Combination of equations (1-5) and (1-6) gives a value for $d\rho$:

$$d\rho = \frac{4.6 \sin \psi_2}{\mu} \qquad (7\text{-}2)$$

Substituting for d in equation (7-1) gives

$$m_a = \frac{4.6 \sin \psi_2}{\mu} \cdot \pi r^2 \qquad (7\text{-}3)$$

Using a value of 35° for ψ_2 and 1.5 cm for r, the approximate value for m_a is given as

$$m_a = \frac{17.5}{\mu} \quad \text{(g)} \qquad (7\text{-}4)$$

Table 7-2. Penetration Depths and Amounts of Specimen Contributing to Measured Fluorescence Intensity

Analyte α lines	λ (Å)	E (keV)	Matrix	Mass Absorption Coeff. (cm²/g)	Penetration Depth d	Analyzed Mass (mg)
C K	44.0	0.28	Steel	15,000	0.5 μm	0.001
Mg K	9.89	1.24	Cement	2,000	5 μm	0.009
U M	3.91	3.17	Sandstone	500	20 μm	0.035
Ca K	3.36	3.69	Cement	400	25 μm	0.044
Ba L	2.78	4.47	Rock	150	65 μm	0.114
Fe K	1.94	6.40	Steel	120	35 μm	0.147
Lu L	1.62	7.66	Rock	70	140 μm	2.450
Pb L	1.18	10.5	Gasoline	3	1.1 μm	6.930
U L	0.91	13.6	Sandstone	25	400 μm	0.700
Mo K	0.71	17.4	Steel	35	120 μm	0.504
Ba K	0.39	32.0	Rock	2	0.5 μm	8.750
Lu K	0.23	53.6	Rock	1	1.0 μm	17.5

Table 7-2 lists typical values for various analyte lines in different matrices, that have been calculated using equation (7-4).

7.2 DIRECT ANALYSIS OF SOLID SAMPLES

While the direct analysis of the sample in the spectrometer would appear to offer significant advantages both in speed and avoidance of contamination in the specimen preparation process, in point of fact, direct analysis is only possible on special occasions. As an example, it is sometimes possible to compress small sample chips into a briquet by high pressure and analyze directly. Although this may give a somewhat uneven surface, it is possible to correct for that by ratioing analytical line intensities to another major sample element line [1].

By far the greatest potential problem in the direct analysis of bulk solids and powders is that of local heterogeneity. As was shown in section 7.1, the actual penetration of X-rays into the specimen is generally rather small and is frequently in the range of a few microns. Where the specimen is heterogeneous over the same range as the penetration depth, what the spectrometer actually analyzes may not be representative of the whole specimen. As in example, the data in table 7-2 represent a series of various elements in different matrices representing a wide range in mass absorption coefficients.

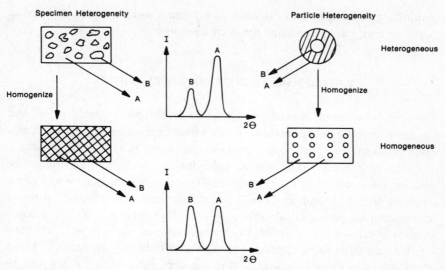

Figure 7-2. Effect of specimen heterogeneity and particle heterogeneity on relative X-ray line intensities.

It will be seen that for the lower atomic number elements, penetration depths may be of the order of only a few microns. The problem in specimen preparation is then to ensure that the relatively thin surface layer actually analyzed is truly representative of the bulk of the sample. This problem may manifest itself either in terms of specimen heterogeneity or particle heterogeneity, as illustrated in figure 7-2. This represents the measurements of two elements A and B, first in a heterogeneous specimen, then in a heterogeneous particle. In the case of specimen heterogeneity, if the penetration of the X-ray beam is of the same order as the particle size, the measured intensity ratio of lines from the two phases A and B is quite different to that which would be obtained from a homogeneous sample. A similar effect may be seen if individual particles are hetergeneous. Such an effect can occur, for example, in the case of partially oxidized pyrite (FeS_2) and chalkopyrite ($CuFeS_2$) minerals.

If the penetrating power of the characteristic radiation is sufficiently large, it may be possible to analyze certain materials directly. This is particularly important for the analysis of living matter. As an example, measurements have been carried out to determine the amount of in vivo tibia lead using a [109]Cd radioisotope source spectrometer [2]. Such a technique allows rapid monitoring of workers with chronic lead exposure, and it has been found that this is an acceptable technique for such screening [3].

7.3   PREPARATION OF POWDER SAMPLES

The most common method of preparing powder samples is first to grind, and then to pelletize at high pressure. While grinding is an extremely quick and effective means of reducing the particle size, there is always the potential problem of contaminating the sample during the grinding process. In practice, some materials may be sufficiently soft and homogeneous to allow direct pelletization and analysis. Such a technique is applicable to many pharmaceutical products [4], although it is often necessary to add a small amount of cellulose as a binder. In general, however, the sample is too hard for this technique to be applied, and means of reducing the particles to an acceptable size must be sought. Many studies have been carried out to quantify the magnitude of this problem-not just in X-ray fluorescence, but also in other spectroscopic disciplines (e.g. ref. 5). Probably the most commonly employed grinding device is the disk mill (shatterbox). This device consists of a series of concentric rings, plus an inner solid disk, that are shaken back and forth very vigorously. The actual sample container may be made of hardened steel, agate or tungsten carbide. These mills are very efficient and are able to reduce powders to less than 325 mesh in a matter of minutes. Unfortunately, they can also be a source of contamination. As an example, Tuff [6] has reported a detailed study of the contamination of a quartzite cobble (99.5% quartz and 0.5% alumina) during crushing and grinding. He reported major pickup of iron, manganese, cobalt and chromium from the steel jaw crusher.

Probably the most effective means of preparing a homogeneous powder sample is the borax fusion method. This method was first proposed by Claisse in 1957 [7]; its principle involves fusion of the sample with an excess of sodium or lithium tetraborate and casting into a solid bead. Chemical reaction in the melt converts the phases present in the sample into glasslike borates, giving a homogeneous bead of controllable dimensions that is ideal for direct placement in the spectrometer. [For a detailed description of the principles of the borax fusion method the reader is referred to *Quantitative Spectrometry* by R. Jenkins, R. W. Gould, and D. A. Gedcke (Marcel Dekker, 1981), Chapter 7, "Specimen preparation".] While manual application of this technique is rather time consuming, a number of automated and semiautomated borax bead making machines are commercially available (e.g. refs. 8, 9) and these devices are able to produce multiple samples in a matter of minutes. The ratio of sample to fusion mixture is quite critical, since this will not only determine the speed and degree of completion of the chemical reaction, but also the final mass absorption coefficient of the analyzed bead and the actual dilution factor applied to the analyte elements. Some control is possible over these factors, by use of fusion aids (such as iodides and peroxides) and use of high atomic number absorbers as part of the fusion mixture (barium and lanthanum salts have been employed for this purpose). Quantitative comparisons of the variables is not easy because of synergistic effects. Claisse's original method proposed a sample to sodium tetraborate ratio of 100 : 1, and in 1962 Rose et al. [10] suggested the use of lithium tetraborate in the ratio of 4 : 1. The same authors also suggested use of lanthanum oxide as a heavy absorber. Some years later Norrish and Hutton [11] concluded that for whole rock analysis a ratio of 5.4 : 1 was ideal, along with added lithium carbonate to lower the fusion temperature. More recently, Bower and Valentine [12] have published a critical review of these various techniques as applied to whole rock analysis, and have compared the results obtained from different recipes with similar data obtained from pressed pellets.

7.4   DIRECT ANALYSIS OF SOLUTIONS

The detection limits directly achievable by the X-ray fluorescence method generally lie in the low ppm range, or in absolute terms, a nominal detection limit of 0.1 to 0.01 μg on a thin film sample. In favorable cases, such as the transition elements, this limit may go to about 0.1 ppm, and in the most unfavorable cases, such as the lower atomic number elements, to about 50 ppm. One of the special problems encountered in the direct analysis of (for example) water samples stems from the need to support the specimen under examination. Most conventional spectrometers irradiate the sample from below, and the support film attenuates the signal from the longer wavelength characteristic lines as well as introducing a significant blank. Absorption by air also becomes an important factor for the measurement of wavelengths longer than about 2Å and the need to work in a helium atmosphere introduces further attenuation. Thus, in almost all cases, some preconcentration technique is applied to the water sample before analysis. Although the sensitivity obtainable is barely sufficient for direct analysis, a wide range of preconcentration techniques has been developed that bring the concentrations of the required elements well within the range of the system. These preconcentration methods are sufficiently well developed that they do not compromise the inherent speed and accuracy of the X-ray fluorescence method. The absence of geometric constraints in the case of the energy dispersive spectrometer makes it ideal for the development of special dedicated instruments for trace analysis, and a useful outgrowth of this has been the use of particle excited spectrometry (PIXE). This technique has already been discussed in section 5.6. As was mentioned, this technique offers tremendous potential for the trace analysis of limited amounts of material.

7.5   ANALYSIS OF SMALL AMOUNTS OF SAMPLE

The irradiation area in a typical X-ray spectrometer is of the order of 5 cm^2, and the penetration depth of an average wavelength about 20 μm. This means that even though 20 g of sample may be placed in the spectrometer, the analyzed volume is still only of the order of 50 mg. The smallest sample that will give a measurable signal above background is at least three orders of magnitude less than this mass, so the lowest analyzable sample is of the order of 0.05 mg, provided that the sample is spread over the full irradiation area of the spectrometer sample cup. Where this is impracticable, the smallest analyzable value is increased by roughly the total irradiation area of the cup (about 5 cm^2) over the actual area of the sample that the primary X-ray beam sees.

A second important area of materials analysis involving small samples involves the investigation of thin films. Although the technique is limited to rather large area, typically a few square millimeters, it does provide useful information about bulk composition of surface films.

In the analysis of even large masses of material, it can be easily shown that the actual quantity analyzed is very small. Since penetration depths are of the order of a few tens of microns, and since the irradiation area is typically a few square centimeters, the actual weight of sample analyzed is only of the order of a few milligrams. Table 7-2 lists some examples of typical samples, and it will be seen that the actual mass of sample analyzed can range from a few milligrams to several grams, depending on the actual analytical situation. A consequence of this is that when analyzing small amounts of material, it is far better to spread the sample out over the irradiated area, than to make a special limited area sample holder. In fact, excellent sensitivity can be obtained even on milligram samples. A good example of this is found in the analysis of air pollutants collected on filter paper.

It is interesting to compare data obtained with conventional X-ray spectrometers with those from direct excitation X-ray analysis using electrons—as in electron probe microanalysis (EPMA), or proton induced X-ray emission (PIXE). Using conventional sources, wavelength and energy dispersive spectrometers can achieve MDL's in the 1 to 10 ng/cm^2 range [13]. However, these instruments use an irradiated area of several square centimeters, so the absolute mass being detected is of the order of 10^{-7}–10^{-9} g. The EMPA method can achieve MDL's of the order of 10^{-14} g [14], and the PIXE method about 10^{-10} g [15].

7.6 PRECONCENTRATION TECHNIQUES

While the direct analysis of a solution offers an ideal sample for analysis, all too often the concentrations of the analyte elements are too low to give an adequate signal above background. In these cases preconcentration techniques must be used to bring the analyte concentration within the sensitivity range of the spectrometer. In principle, one could easily accomplish this simply by evaporation. In order to achieve detection limits at the low ppb level, this would mean the evaporation of about 100 ml of water. Although it is not theoretically necessary to take the sample completely to dryness, it is much more convenient to do so because, by this means, one can obtain a handleable specimen. Unfortunately, taking the sample to dryness does cause experimental problems due to fractional crystallization, splashing of the sample as it reaches dryness, and so on. For these practical reasons preconcentration by evaporation has not found a great deal of application. On the other hand, the application of evaporation preconcentration techniques does show some promise when combined with special techniques for reducing the relatively high inherent background observed in classical X-ray fluorescence methods. The method of TRXRF, discussed in section 5.4, utilizes the total reflection of X-rays from a highly polished surface. Aiginger and his co-workers [16] have achieved sensitivities down to the ppb level by evaporation of small (about 5 μL) samples of water onto a very flat quartz-glass plates. A thin layer of insulin was used to give a good distribution of the evaporated sample across the surface of the optical flat, and an energy dispersive spectrometer was used to measure and analyze the characteristic X-ray emission.

Many different chemical and physical preconcentration techniques have been proposed, including include surface adsorption on activated carbon [17], electrodeposition [18], precipitation chromatography [19], liquid–liquid extraction [20], immobilized reagents [21], preconcentration on ion exchange treated polyurethane foam [22], plus a variety of other techniques well known to the analytical chemist. However, by far the most popular, and also the most successful, techniques have been those based on the use of ion exchange resins. The major advantage of most ion exchange methods is that the functional group is immobilized on a solid substrate, providing the potential to batch extract ions from solution. The sample itself can be either the actual exchanged resin or a separate sample containing the appropriate ions re-eluted from the resin. The success of the method depends to a large extent on the recovery efficiency of the resin, which in turn is determined by the affinity of the ion exchange material for the ions in question and the stability of complexes present in solution. Preconcentration factors of up to 4×10^4 can be achieved from suitable ion exchange material using around 100 mg of resin [23].

One of the most useful ion exchange resins is Chelex-100. This contains iminodiacetic functional groups and acts chemically very similarly to EDTA (ethylenediamine tetraacetic acid). In shows a good recovery efficiency for a wide range of ions, and its chemistry can be predicted from prior experiments with EDTA. Against this, however, it is not very successful in the separation of ions from solution that are high in iron and calcium (e.g. sea water), which elements occupy the available sites to the exclusion of the trace element ions. Many other ion exchange resins have also been used; for examples, Dowex 1X8 has been employed for the determination of cobalt in iron rich materials [24], and Wolfatit RO resin for the separation of trace amounts of gold in cyanide liquors [25]. It is sometimes found convenient to use filter paper impregnated with ion exchange resin. This technique has been found especially useful in the extraction of low concentrations of rare earth elements [26]. The exchange resin can also be employed as a membrane through which the solution to be analyzed is passed [27]. This is a particularly convenient means of separation for the X-ray fluorescence method, because the paper can be mounted directly in the spectrometer for analysis following the separation process. As an example of the use of Chelex-100, 200 mL of water were passed through two Chelex-100 membranes for a period of about 20 minutes. The enrichment factors obtained were in excess of 1000, allowing the separation of potassium, calcium, manganese, cobalt, nickel, copper, zinc, rubidium and strontium as chlorides or nitrates at element concentrations in the range 10 ppm to 10 ppb [28].

Another method that has been employed with some success, especially to preconcentrate elements in natural water samples, is the use of coprecipitation. This relatively simple method offers the advantage of giving a fairly uniform deposit that can be easily collected. One of the earlier applications of this technique involved the use of iron hydroxide as a coprecipitant for the determination of iron, zinc and lead in surface waters [29]. One of the more popular coprecipitants in use today is ammonium pyrrolidine dithiocarbamate (APDC). In the application of this method to the analysis of natural waters, detection limits in the range 0.4 to 1.2 ppb have been claimed for the elements vanadium, zinc, arsenic, mercury and lead [30]. Other coprecipitants have been described including the use of iron dibenzyl dithiocarbamate for the determination of uranium at the ppb level in natural waters [31], zirconium dioxide for the determination of arsenic in river water [32], and polyvinyl pyrrolidone-thionalide for the determination of iron, copper, zinc, selenium, cadmium, tellurium, mercury and lead in waste and natural water samples [33].

REFERENCES

1. Tokuda, T., et al. (1985). *R & D, Res. Dev.* (Kobe Steel Ltd.) **35**, 8.
2. Somerville, L. J., Chettle, D. R., and Scott, M. C. (1985). *Phys. Med. Biol.* **30**, 929.
4. Sanner, G., and Usbeck, H. (1985). Pharmazie **40**, 544.
5. Thompson, G., and Bankston, D. C. (1970). *Appl. Spectrosc.* **24**, 210.
6. Tuff, M. A. (1985). Adv. X-Ray Anal. **29**, 565.
7. Claisse, F. (1957). *Norelco Reporter* **4**, 3.
8. Claisse Fusion Stirrer Device, Spex Industries, Metuchen, New Jersey.
9. Willay, G. (1986). "'Perl-X' and its derivatives in mineral analysis", *Cah. Inf. Tech./Rev. Metall* **83**, 159.
10. Rose, H., Adler, I., and Flanagan, F. J. (1960). *U.S. Geol. Survey Prof. Pap.* **450-B**, 80.
11. Norrish, K., and Hutton, J. (1969). *Geochim. Cosmochim. Acta* **33**, 432.
12. Bower, N., and Valentine, G. (1986). *X-Ray Spectrom.* **15**, 73.
13. Birks, L. S., and Gilfrich, J. V. (1978). *Appl. Spectrosc.* **32**, 204.
14. Gilfrich, J. V., et al. (1982). *Adv. X-Ray Anal.* **26**, 313.

15. Johansson, E. M., and Akselsson, K. R. (1981). *Nucl. Instrum. Methods* **181**(1–3), 221–226.

16. Wobrauschek, P., and Aiginger, H. (1975). *Anal. Chem.* **47**, 852–855.

17. Howe, P. T. (1980). At. Energy Can. Ltd., AECL-6444, 11 pp.

18. Vassos, B. H., Hirsch, R. F., and Letterman (1973). H., *Anal. Chem.* **45**, 792.

19. Zeronsa, W. P., Dobkowski, G., and Siggia, S. (1974). *Anal. Chem.* **46**, 309.

20. Leyden, D., et al. (1980). *Pergamon Ser. Environ. Sci. (Anal. Tech. Environ. Chem.)* **3**, 469.

21. Hercules, D. M., et al. (1973). *Anal. Chem.* **45**, 1973.

22. Torok, S., Van Dyke, P., and Van Grieken, R. (1986). *X-Ray Spectrom.* **15**, 7.

23. Leyden, D. E., Patterson, T. A., and Alberts, J. J. (1975). *Anal. Chem.* **47**, 733.

24. Roelands, I. (1985). *Chem. Geol.* **51**, 3.

25. Peter, H. J., Braun, J., Dietze, U., and Volke (1985). P., *Z. Chem.* **25**, 374.

26. An, Q. (1985). *New Front. Rare Earth Sci. Appl. Proc. Int. Conf. Rare Earth Dev. Appl.* **1**, 551.

27. Campbell, W., Spano, E. F., and Green, T. E. (1966). *Anal. Chem.* **38**, 987.

28. Van Grieken, R. E., Bresele, C. M., and van der Goe, B. M. (1977). *Anal. Chem.* **49**(9), 1326.

29. Bruninx, E., and van Meijl, E. (1975). *Anal. Chim. Acta* **80**, 85.

30. Pradzynski, A. H., Henry, R. E., and Stewart, J. S. (1975). *Radiochem. Rad.* **21**(5), 273.

31. Caravajal, G. S., Mahan, K. J., and Leyden, D. E. (1982). *Anal. Chim. Acta* **135**, 205.

32. Katsuno, T. (1981). *Nagano-ken Eisei Kogai Kenkyusho Kenkyu Hokoku* **3**, 52.

33. Panayappan, R., et al. (1978). *Anal. Chem.* **50**, 1125.

CHAPTER

8

USE OF X-RAY SPECTROMETRY FOR QUALITATIVE ANALYSIS

8.1 INTRODUCTION TO QUALITATIVE ANALYSIS

Both the simultaneous wavelength dispersive spectrometer and the energy dispersive spectrometers lend themselves admirably to the qualitative analysis of materials. As was shown in equation (1-1), there is a simple relationship between the wavelength or energy of a characteristic X-ray photon and the atomic number of the element from which the characteristic emission line occurs. It has also been shown that each element will emit a number of characteristic lines within a given series (K, L, M, etc.). Thus by measuring the wavelengths, or energies, of a given series of lines from an unknown material the atomic numbers of the excited elements can be established.

Because the characteristic X-ray spectra are so simple, the actual process of allocating atomic numbers to the emission lines is a relatively easy process and the chance of making a gross error is rather small. In comparison, the procedures for the qualitative analysis of multiphase materials with the X-ray powder diffractometer is much more complex. There are only 100 or so elements, and within the range of the conventional spectrometer each element gives, on an average, only half a dozen lines. In X-ray diffraction, on the other hand, there are perhaps as many as several million possible compounds, each of which can give on an average 50 or so lines. Similarly if one compares the X-ray emission spectrum with the ultraviolet emission spectrum: Since the X-ray spectrum arises from a limited number of inner orbital transitions, the number is X-ray lines is similarly rather few. Ultraviolet spectra, on the other hand, arise from transitions to empty levels—of which there may be many, leading to a significant number of lines in the UV emission spectrum. A further benefit of the X-ray emission spectrum for qualitative analysis is that because transitions do arise from inner orbitals, the effect of chemical combination, i.e. valence, is almost negligible.

8.2   RELATIONSHIP BETWEEN WAVELENGTH AND ATOMIC NUMBER

In section 1.3 it was shown that one of the ways an excited atom can revert to its original ground state is by transferring an electron from an outer atomic level to fill the vacancy in the inner shell. An X-ray photon is then emitted from the atom, having an energy equal to the energy difference between the initial and final states of the transferred electron. As an example, the binding energy of the K level in iron is about 7,100 eV, and that for the L level about 700 eV. Thus, when the atom moves from the K^+ state to the L^+ state, there will be a total energy change of $7,100 - 700 = 6,400$ eV. In practice, millions of atoms are involved in the excitation of a given specimen, and all possible deexcitation routes are taken. It was also indicated that the various deexcitation routes can be defined by a simple set of selection rules that account for the majority of the observed wavelengths. The classically accepted nomenclature system for the observed lines is that proposed by Siegbahn in the 1920's, and is referred to as the *Siegbahn system*. The wavelengths that fit the selection rules are called *normal* or *diagram* lines.

Figure 8-1 shows Moseley diagrams for the K, L and M series. In order to simplify the figure, only the strongest lines in each series have been shown. These diagrams are plots of the reciprocal of the square root of the wavelength as a function of atomic number. As indicated by Moseley's law, shown in the figure, such plots should be linear. A scale directly in wavelength is also shown, to indicate the range of wavelengths over which a given series occurs.

Each electronic state in an atom can be defined by four quantum numbers. The first is the *principal quantum number n*, which can take on all integral values. When *n* is equal to 1, the level is referred to as the K level; when *n* is 2, the L level and so on. The second is the *orbital angular momentum quantum number l*, and this can take all values from $n - 1$ to zero. The third is the *magnetic quantum number m*, which can take all values from $+l$ through zero to $-l$. The fourth is the *spin quantum number s*, which has a value of $\pm\frac{1}{2}$. Note that the total angular momentum quantum number of an electron is given by the vector sum *l* and *s*. This sum is denoted by J. The Pauli exclusion

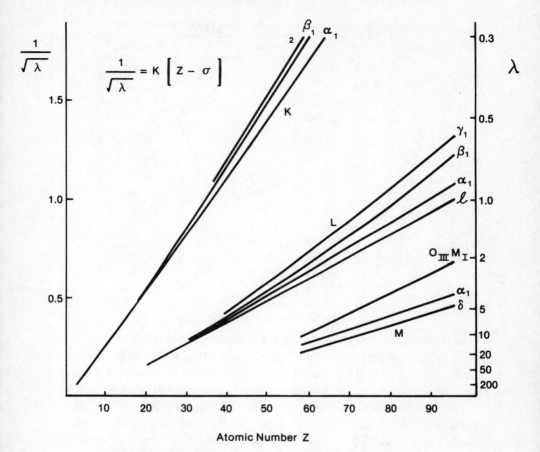

Figure 8-1. Moseley diagrams for the K, L and m series.

principle states that no two electrons within a given atom can have the same set of quantum numbers, and so, as electrons are added, a series of levels or *shells* is built up.

Table 8-1 gives the atomic structures of the first three principal shells. The first shall, the K shell, has a maximum of two electrons, and these are both in the 1s level. Since the value of J must be positive in this instance, the only allowed value is $\frac{1}{2}$. In the second shell, the L shell, there are eight electrons: two in the 2s level and six in the 2p levels. In this instance J has a value of $\frac{1}{2}$ for the 1s level and $\frac{3}{2}$ or $\frac{1}{2}$ for the 2p level. Thus giving a total of three possible L transition levels. These are referred to as L_I, L_{II} and L_{III} respectively. In the

Table 8-1. Atomic Structures of the First Three Principal Shells

Shell (Electrons)	n	l	m	s	Orbitals	J
K (2)	1	0	0	$+\frac{1}{2}$	1s	$\frac{1}{2}$
L (8)	2	0	0	$+\frac{1}{2}$	2s	$\frac{1}{2}$
	2	1	1	$+\frac{1}{2}$		
	2	1	0	$+\frac{1}{2}$	2p	$\frac{1}{2}, \frac{3}{2}$
	2	1	-1	$+\frac{1}{2}$		
M (18)	3	0	0	$+\frac{1}{2}$	3s	$\frac{1}{2}$
	3	1	1	$+\frac{1}{2}$		
	3	1	0	$+\frac{1}{2}$	3p	$\frac{1}{2}, \frac{3}{2}$
	3	1	-1	$+\frac{1}{2}$		
	3	2	2	$+\frac{1}{2}$		
	3	2	1	$+\frac{1}{2}$		
	3	2	0	$+\frac{1}{2}$	3d	$\frac{3}{2}, \frac{5}{2}$
	3	2	-1	$+\frac{1}{2}$		
	3	2	-2	$+\frac{1}{2}$		

M level, there are a maximum of eighteen electrons: two in the 3s level, eight in the 3p level and ten in the 3d level. Again, with the values of $\frac{3}{2}$ or $\frac{1}{2}$ for J in the 3p level, and $\frac{5}{2}$ and $\frac{3}{2}$ in the 3d level, a total of five M transition levels are possible. Similar rules can be used to build up additional levels, N, O, etc.

The selection rules for the production of diagram lines state that the principal quantum number must change by at least one, the orbital quantum number must change by only one, and the quantum number J must change by zero or one. Transition groups may now be constructed as illustrated in table 8-2, based on the appropriate number of transition levels. Application of the selection rules indicates that in, for example, the K series, only L_{II} to K and L_{III} to K transitions are allowed for a change in the principal quantum number of one. There are an equivalent pairs of transitions for $n = 2$, $n = 3$, $n = 4$, etc.

Figure 8-2 shows the lines that are observed in the K series. Three groups of lines are indicated. The normal lines are shown on the left hand side of the figure. These consist of three pairs of lines from the L_{II}/L_{III}, M_{II}/M_{III} and N_{II}/N_{III} subshells respectively. While most of the observed fluorescent lines are normal, certain lines may also occur in X-ray spectra that do not at first sight fit the basic selection rules. These lines are called *forbidden* lines and are shown in the center portion of the figure. Forbidden lines typically arise from

Table 8-2. Construction of Transition Groups

Transition Group	n	l	J
K	1	0	$\frac{1}{2}$
L_I	2	0	$\frac{1}{2}$
L_{II}	2	1	$\frac{1}{2}$
L_{III}	2	1	$\frac{3}{2}$
M_I	3	0	$\frac{1}{2}$
M_{II}	3	1	$\frac{1}{2}$
M_{III}	3	2	$\frac{3}{2}$
M_{IV}	3	1	$\frac{3}{2}$
M_V	3	2	$\frac{5}{2}$
N_I	4	0	$\frac{1}{2}$
N_{II}	4	1	$\frac{1}{2}$
N_{III}	4	1	$\frac{3}{2}$
N_{IV}	4	2	$\frac{3}{2}$
N_V	4	2	$\frac{5}{2}$
N_{VI}	4	3	$\frac{5}{2}$
N_{VII}	4	3	$\frac{7}{2}$

outer orbital levels where there is no sharp energy distinction between orbitals. As an example, in the transition elements, where the 3d level is only partially filled and is energetically similar to the 3p levels, a weak forbidden transition (the α_5) is observed. A third type of lines may also occur, called *satellite lines*, which arise from dual ionizations. Following the ejection of the initial electron in the photoelectric process, a short, but finite, period of time elapses before the vacancy is filled. This time period is called the lifetime of the excited state. For the lower atomic number elements, this lifetime increases to such an extent that there is an significant probability that a second electron can be ejected from the atom before the first vacancy is filled. The loss of the second electron modifies the energies of the electrons in the surrounding subshells, and other pairs of X-ray emission lines are produced, corresponding to the $\alpha_1 \alpha_2$. In the K series the most common of these satellite lines are the $\alpha_3 \alpha_4$ and the $\alpha_5 \alpha_6$ doublets. These lines are shown at the right hand side of the figure. Although, because they are relatively weak, neither forbidden transitions nor satellite lines have great analytical significance, they may cause some confusion in qualitative interpretation of spectra and may even be misinterpreted as coming from trace elements.

In practice, the number of lines observed from a given element will depend

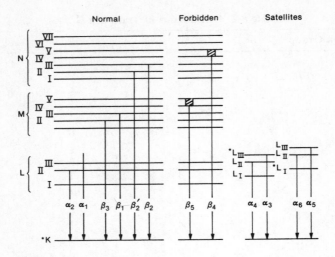

Figure 8-2. Normal, forbidden and satellite lines in the K series.

upon the atomic number of the element, the excitation conditions and the wavelength range of the spectrometer employed. Generally, commercial spectrometers cover the range 0.3 to 20 Å (newer instruments may allow measurements in excess of 100 Å), and three X-ray series are covered by this range: the K series, the L series and the M series, corresponding to transitions to K, L and M levels respectively. Each series consists of a number of groups of lines. The strongest group of lines in the series is denoted α, the next strongest β, and the third γ. There are a much larger number of lines in the higher series, and for a detailed list of all of the reported wavelengths the reader is referred to the work of Bearden [1].

In X-ray spectrometry most of the analytical work is carried out using either the K or the L series wavelengths. The M series may, however, also be useful, especially in the measurement of higher atomic numbers. One practical problem in working with the M series in that many of the lines have not been given names within the Siegbahn system. In recent years the International Union of Pure and Applied Chemistry has been addressing the problems of spectroscopic nomenclature and as part of this project has suggested an alternative nomenclature system based on edge designation [2]. This system is referred to as the IUPAC notation. Table 8-3 compares the Siegbahn notation and the IUPAC notation for the strongest lines in the K, L and M series.

Table 8-3. Correspondence Between Siegbahn and IUPAC Notation for Diagram Lines

Siegbahn	IUPAC	Siegbahn	IUPAC	Siegbahn	IUPAC
$K\alpha_1$	$K\text{-}L_3$	$L\alpha_1$	$L_3\text{-}M_5$	$L\gamma_1$	$L_2\text{-}N_4$
$K\alpha_2$	$K\text{-}L_2$	$L\alpha_2$	$L_3\text{-}M_4$	$L\gamma_2$	$L_1\text{-}N_2$
$K\beta_1$	$K\text{-}M_3$	$L\beta_1$	$L_2\text{-}M_4$	$L\gamma_3$	$L_1\text{-}N_3$
$K\beta_2^{I}$	$K\text{-}N_3$	$L\beta_2$	$L_3\text{-}N_5$	$L\gamma_4$	$L_1\text{-}O_3$
$K\beta_2^{II}$	$K\text{-}N_2$	$L\beta_3$	$L_1\text{-}M_3$	$L\gamma_4'$	$L_1\text{-}O_2$
$K\beta_3$	$K\text{-}M_2$	$L\beta_4$	$L_1\text{-}M_2$	$L\gamma_5$	$L_2\text{-}N_1$
$K\beta_4^{I}$	$K\text{-}N_5$	$L\beta_5$	$L_3\text{-}O_{4,5}$	$L\gamma_6$	$L_2\text{-}O_4$
$K\beta_4^{II}$	$K\text{-}N_4$	$L\beta_6$	$L_3\text{-}N_1$	$L\gamma_8$	$L_2\text{-}O_1$
$K\beta_{4x}$	$K\text{-}N_4$	$L\beta_7$	$L_3\text{-}O_1$	$L\gamma_8'$	$L_2\text{-}N_{6\,(7)}$
$K\beta_5^{I}$	$K\text{-}M_5$	$L\beta_7$	$L_3\text{-}N_{6,7}$	$L\eta$	$L_2\text{-}M_I$
$K\beta_5^{II}$	$K\text{-}M_4$	$L\beta_9$	$L_1\text{-}M_5$	$L\ell$	$L_3\text{-}M_1$
		$L\beta_{10}$	$L_1\text{-}M_4$	Ls	$L_3\text{-}M_3$
		$L\beta_{15}$	$L_3\text{-}N_4$	Lt	$L_3\text{-}M_2$
		$L\beta_{17}$	$L_2\text{-}M_3$	Lu	$L_3\text{-}N_{6,7}$
				Lv	$L_2\text{-}N_{6\,(7)}$

8.3 RELATIVE INTENSITIES OF X-RAY LINES

Figure 8-3 shows the strongest lines in the K, L and M series for the element tungsten ($Z = 74$). It will be seen tha the K series is dominated by the $\alpha_1\,\alpha_2$ doublet, with a group of β lines at the short wavelength side of the α's. The L series contains three main groups of lines: the α's, the β's and the γ's. The strongest line is the α_1, and the second strongest the β_1. The M series is dominated by the unresolved $\alpha_1\,\alpha_2$ doublet, with the β line the next strongest.

While the intensities observed for tungsten are typical of most of the periodic table, there are significant differences in the intensities at the extreme atomic number limits of the various series. As an example, in the L series, both the β_2 and the γ_1 arise from the 4d levels. Since elements of atomic number less than 39 do not have electrons in these levels, these two lines (among others) do not appear in the lower atomic number L spectra. By far the greatest differences in spectral intensities appear in the low atomic

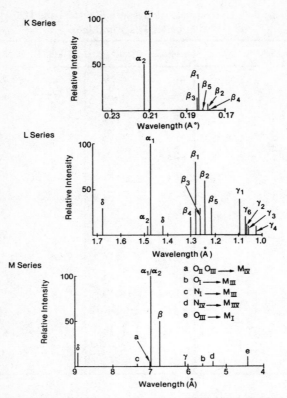

Figure 8-3. Principal lines in the K, L and M series for tungsten.

number K spectra. Figure 8-4 shows the K spectra for four elements—tin ($Z = 50$), zinc ($Z = 30$), calcium ($Z = 20$) and aluminum ($Z = 13$). The tin spectrum is typical of the higher atomic number K series and shows a strong, resolved α_1 α_2 doublet; a relatively strong β_1 β_3 doublet that masks the forbidden β_5, and a weaker unresolved β_2 β_2' doublet that masks the forbidden β_4. Moving 20 atomic numbers down the periodic table to zinc leads to the disappearance of the forbidden β_4. The α_1 α_2 doublet is now only partly resolved, and the β_1 β_3 doublet is somewhat weaker than in the case of tin. Moving another 10 atomic numbers down to calcium leads to a further weakening of the β_1 β_3 doublet and the complete disappearance of the forbidden β_5. However, now we see the appearance of the α_3 α_4 and α_5 α_6 satellite doublets. Moving a further 7 atomic numbers down to aluminum

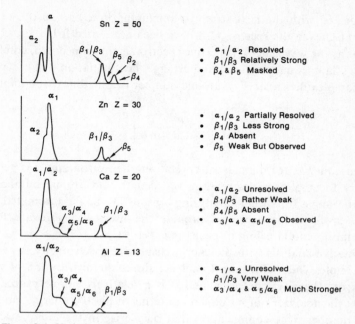

Figure 8-4. Variations in K emission spectra as a function of atomic number.

leads to satellites of greater intensity and an even weaker β_1 β_3 doublet. At very low atomic numbers the transferred electrons almost all come from valence bands, and there is an abrupt transition from reasonably sharp line spectra to band spectra below about atomic number 9. Similar changes in the intensity distribution of spectral lines occur in the longer wavelength L and M series lines.

8.4 QUALITATIVE INTERPRETATION OF X-RAY SPECTROGRAMS

Qualitative analysis with the X-ray spectrometer involves identifying each line in the measured spectrogram. As will be apparent from the typical intensity distributions shown in figure 8-3, a preliminary visual study of the spectrogram will generally indicate to which line series a given group of lines belongs. The strongest line is generally assumed to be a Kα, or Lα, etc. After this strongest line has been assigned to a certain element, all other lines from this element are checked off against a set of tables (e.g. ref. 3). The process is then repeated with the next strongest remaining line, and so on, until all significant lines in the spectrogram have been accounted for.

As far as the wavelength dispersive spectrometer is concerned, combining Moseley's law [equation (1-1)] and Bragg's law [equation (1-8)] reveals a rather complex dependence of atomic number on the diffraction angle, viz.

$$Z = \left[\frac{1}{K} \cdot \frac{n}{2d} \cdot \frac{1}{\sin \theta} \right]^{1/2} \tag{8-1}$$

As a consequence, for manual interpretation of spectrograms it is generally necessary to employ tables relating wavelength and atomic number with diffraction angle for specific analyzing crystals [3]. Some automated wavelength dispersive spectrometers provide the user with software for the interpretation and labelling of peaks (e.g. ref. 4).

Qualitative analysis with the energy dispersive spectrometer is generally rather simple. The multichannel analyzer stores the distribution of voltage pulses from the Si(Li) detector, and it is a relatively simple procedure to calibrate the analyzer output directly in terms of photon energy and thence atomic number. One complication that may arise in the interpretation of spectra recorded with the energy dispersive spectrometer is the appearance of artifacts in the acquired spectra. The artifacts may include "sum peaks," "escape peaks" and diffraction lines [5]. However, such phenomena are generally recognized by the user, and do not present undue difficulty to the experienced operator.

8.5 SEMIQUANTITATIVE ANALYSIS

The X-ray fluorescence method is a particularly useful tool for semiquantitative analysis. A special advantage offered by the method is that the sensitivity does not change dramatically from element to element, but follows a rather smooth curve, with discontinuities where instrument parameters are changed. The lower portion of figure 8-5 shows the line series covered over the range 0.3 to 200 Å. The middle portion of the figure illustrates typical choices of detector, analyzing crystal and collimator. Within a given set of these parameters the sensitivity of the spectrometer is determined mainly by two factors—the fluorescent yield for the excited wavelength, and the primary X-ray source overvoltage. As was shown in equation (4-1), provided that the tube current is fixed, the intensity of a given excited X-ray line is

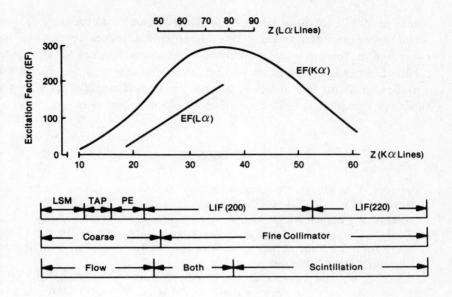

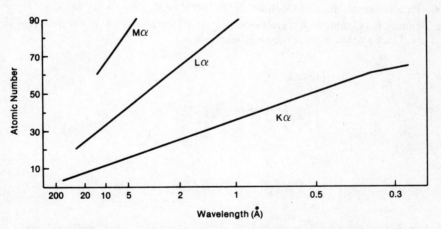

Figure 8-5. Wavelength range and sensitivity of the wavelength dispersive spectrometer.

proportional to $(V_0 - V_c)^{1.6}$. This quantity is referred to as the overvoltage, and its product with the fluorescent yield is sometimes called the excitation factor. In the upper part of the figure, curves are shown for the excitation factors for the K and L series α lines, as a function of atomic number, for a source at 60 kV. It will be seen that for the K series, this product varies by about three orders of magnitude over the range of measurable elements. The curve is of the form of a bell, peaking at about atomic number 35. The curve for the L series is almost linear between atomic numbers 50 and 90, with a variation of about two orders of magnitude. Use of such curves allows a rapid semiquantitative estimation of composition to be made.

REFERENCES

1. Bearden, J. A. (1964). "X-ray wavelengths," U.S. Atomic Energy Commission Report NYO-10586, 533 pp.
2. Jenkins, R., Manne, R., Robin, J., and Senemaud, C. (1987). *Nomenclature, Symbols, Units and their Usage in Spectrochemical Analysis, Part VIII, Nomenclature System for X-ray Spectroscopy*, International Union of Pure and Applied Chemistry: Oxford England.
3. White, E. W., and Johnson. G. G., Jr. (1970). *X-ray Emission and Absorption Wavelengths and Two-Theta Tables*, ASTM Data Series DS-37A, ASTM: Philadelphia.
4. Garbauskas, M. F., and Goehner, R. P. (1983). *Adv. X-ray Anal.* **26**, 345.
5. Jenkins, R., Gould, R. W., and Gedcke, D. (1981). *Quantitative X-Ray Spectrometry*, Dekker: New York, chapter 8, section 4.

CONSIDERATIONS IN QUANTITATIVE X-RAY FLUORESCENCE ANALYSIS

9.1 CONVERSION OF CHARACTERISTIC LINE INTENSITY TO ANALYTE CONCENTRATION

The great flexibility and range of the various types of X-ray fluorescence spectrometer, coupled with their high sensitivity and good inherent precision, make them ideal for quantitative analysis. In common with all analytical methods, quantitative X-ray fluorescence analysis is subject to a number of random and systematic errors that contribute to the final accuracy of the analytical result. Only by understanding the sources of the errors can they be controlled within reasonable proportions. As in all instrumental methods of analysis, the potentially high precision of X-ray spectrometry can only be translated into high accuracy if the various systematic errors in the analysis process are taken care of.

The precision of a wavelength dispersive system, for the measurement of a single, well-separated line, is typically of the order of 0.1%, and for an energy dispersive system it is about 0.25–0.50%. The major source of the random error is the X-ray source, i.e. the high voltage generator plus the X-ray tube. In addition there is an error arising from the statistics of the actual counting process. The random error can be significantly worse in the case of the energy dispersive system in those cases where full or partial line overlap occurs. Even though good peak and background stripping programs are available to ameliorate this problem, the statistical limitations of dealing with the difference of two large numbers remains. This is probably the biggest hindrance to obtaining precise count data from complex mixtures, and the resulting error can reach several percent in worst cases.

A good rule of thumb that can be used in X-ray fluorescence analysis to estimate the expected standard deviation s at an analyte concentration level C is

$$s = K\sqrt{C + 0.1} \qquad (9\text{-}1)$$

where K varies between 0.005 and 0.05. For example, at a concentration level $C = 25\%$, the expected value of s would be between about 0.025% and 0.25%. A K-value of 0.005 would be considered very high quality analysis, and a value of 0.05 rather poor quality. The value of K actually obtained under routine laboratory conditions depends upon many factors, but with reasonably careful measurements a K-value of around 0.02 to 0.03 can be obtained.

Equation (9-1) is based on the fact that since the production of X-rays is a random process, the standard deviation s for the finite number N of pulses measured is equal to $\sqrt{N}$. The number of pulses measured for a particular experiment is the product of the pulse (counting) rate R and the count time t, with a standard deviation given by:

$$s = \sqrt{Rt} \qquad (9\text{-}2)$$

Provided that all systematic errors are removed, there is a simple straight line relationship between count rate and concentration, i.e. $C = K_1 R$, where K_1 represents the sensitivity of the spectrometer for the line in question. Substituting for R in equation (9-2) gives

$$s = K_2\sqrt{C} \qquad (9\text{-}3)$$

where K_2 is a constant depending on sensitivity and count time.

Equation (9-3) has to be extended somewhat because the counting error is not the only random error involved in the measurement. As was discussed in section 4.2, the short term source error is of the order of 0.1%, so term is added to equation (9-3) to allow for this. Addition of this term gives the form of the equation (9-1).

Table 9-1 lists the four main categories of random and systematic error encountered in X-ray fluorescence analysis. The first category includes the selection and preparation of the sample to be analyzed. Two stages are generally involved before the actual prepared specimen is presented to the spectrometer: sampling and specimen preparation. The actual sampling is rarely under the control of the spectroscopist, and it generally has to be

Table 9-1. Major Sources of Random and Systematic Error

Source	Random	Systematic
1. Sampling	[a]	[a]
Sample preparation	0–1%	0–5%
Sample inhomogeneity	—	0–50%
2. Excitation source	0.05–0.2%	0.05–0.5%
Spectrometer	0.05–0.1%	0.05–0.1%
3. Counting statistics	Time dependent	—
Deadtime losses	—	0–25%
4. Primary absorption	—	0–50%
Secondary absorption	—	0–25%
Enhancement	—	0–15%

[a] Not under analyst's control.

assumed that the container provided does, in fact, contain a representative sample. It will be seen from the table that in addition to a relatively large random error, inadequate sample preparation and residual sample heterogeneity can lead to very large systematic errors. For accurate analysis these errors must be reduced by using a suitable specimen preparation method.

The second category includes errors arising from the X-ray source, and this has already been discussed in some detail in section 4.2. As was stated there, source errors can be reduced to less than 0.1% by use of the ratio counting technique, provided that high frequency transients are absent.

The third category involves the actual counting process. As was shown in section 4.3, systematic errors due to detector dead time may be a problem, but these can be corrected either by use of electronic dead time correctors or by some mathematical approach. There is also a random error associated with the counting process, and this will be discussed in section 9.3.

The fourth category includes all errors arising from interelement effects. Each of the effects listed can give large systematic errors that must be controlled by the calibration and correction scheme. The source of the various interelement effects will be discussed in section 9.4.

9.2 INFLUENCE OF THE BACKGROUND

The background that occurs at a selected characteristic wavelength or energy arises mainly from scattered source radiation. Since scattering increases with decreasing average atomic number of the scatterer, it is found that backgrounds are much higher from low average atomic number specimens than from specimens of high average atomic number. To a first approximation, the background in X-ray fluorescence varies as $1/Z^2$. Since the spectral intensity from the X-ray source increases quite sharply as one approaches a wavelength equal to one-half the minimum wavelength of the continuum, backgrounds from samples excited with bremsstrahlung sources are generally very high at short wavelengths (high energies)—again, especially in the case of low average atomic number samples.

As shown in figure 9-1a, the measured signal is a distribution either of the counting rate R as a function of the angle 2θ (for wavelength dispersive spectrometers), or of the number of counts per channel as a function of energy (energy dispersive spectrometers). A measurement of a line at peak maximum position P gives a peak maximum counting rate R_p. In this instance, where the contribution from the background is insignificant, the analyte concentration is related to R_p. Where the background is significant (figure 9-1b), the measured value of R_p includes a count rate contribution from the background, R_b. The analyte concentration in this case is related to the net counting rate $R_n = R_p - R_b$. Note from the figure that it is not possible to measure R_b directly, and what is generally done is to make a background measurement at a position B that is close to the peak, the assumption being that the background at B is the same as the background under the peak. This assumption will, of course, break down where the peak is superimposed on top of a variable background (figure 9-1c). In this case it is common practice to measure the background on either side of the peak at positions B_1 and B_2, giving counting rates of R_{b1} and R_{b2}. The net intensity R_n is now equal to

$$R_p - \tfrac{1}{2}(R_{b1} + R_{b2}) \tag{9-4}$$

A further complication occurs where the analyte line is partially overlapped by another line (figure 9-1d). Then the measured value of R_p includes a contribution R_b from the background as before, but in addition, a contribution R_b' from the interfering peak. A line overlap correction must be applied, and the assumption is made that the ratio of the peak intensity of the interfering line to the intensity at the interfering shoulder of the line is constant. Thus the net intensity in this case is

$$R_n = R_p - (R_b + k[R_b + kR_0])$$

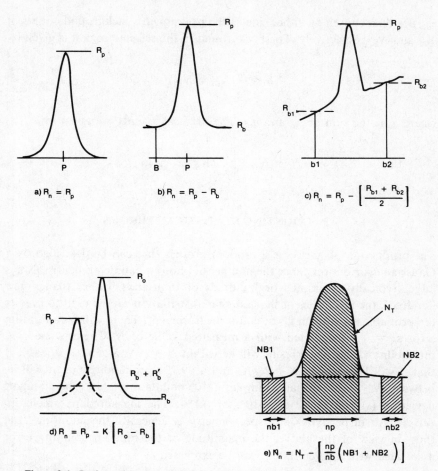

Figure 9-1. Peaks and backgrounds in wavelength and energy dispersive spectrometry.

or

$$R_n \sim R_p - k(R_0 - R_b) \tag{9-5}$$

As illustrated in figure 9-1e, in energy dispersive X-ray fluorescence it is common practice to select a number of channels n_p giving a number of counts N_T, representing a net number of counts N_n on the peak, superimposed on a number of counts B from the background. Two ranges of channels, n_{b1} and

n_{b2}, are then chosen on either side of the peak, giving background counts of N_{B1} and N_{B2} respectively. The background in the selected region is given by

$$B = \frac{n_p}{n_b}(N_{B1} + N_{B2}) \qquad (9\text{-}6)$$

where n_b is the sum of n_{b1} and n_{b2}. The net peak counts are given by

$$N_n = N_T - \left[\frac{n_p}{n_b}(N_{B1} + N_{B2})\right] \qquad (9\text{-}7)$$

9.3   COUNTING STATISTICAL ERRORS

The production of X-rays is a random process that can be described by a Gaussian distribution. Since the number of photons counted is nearly always large (typically thousands or hundreds of thousands, rather than a few hundred), the properties of the Gaussian distribution can be used to predict the probable error for a given count measurement. There will be a random error $s_N = \sqrt{N}$ associated with a measured value of N. There is a 68.3% probability that a given result will lie within $N \pm \sqrt{N}$, a 95.4% probability that it will lie between $N \pm 2\sqrt{N}$, and a 99.7% probability that it will lie between $N \pm 3\sqrt{N}$. As an example, if 10^6 counts are taken, the standard deviation (s) will be $\sqrt{10^6} = 10^3$, or 0.1%. The measured parameter in wavelength dispersive X-ray spectrometry is generally the counting rate; thus, in view of the above, the magnitude of the random counting error associated with a given datum can be expressed as

$$s(\%) = \frac{100}{\sqrt{N}} = \frac{100}{\sqrt{Rt}} \qquad (9\text{-}8)$$

Care must be exercised in relating the counting error (or indeed any intensity related error) with an estimate of the error in terms of concentration. Provided that the sensitivity of the spectrometer in counts/sec per percent is linear, a count error can be directly related to a concentration error. However, where the sensitivity of the spectrometer changes over the range of measured response, a given count error may be much greater when expressed in terms of concentration.

In many analytical situations the peak lies above a significant background, and this adds a further complication to the counting statistical error. An additional factor that must also be considered is that whereas with the scanning wavelength dispersive spectrometer the peaks and background are measured sequentially, in the case of the energy dispersive and the multichannel wavelength dispersive spectrometers, a *single* counting time is selected for the complete experiment; thus all peaks and all backgrounds are counted for the same time. To estimate the net counting error in the case of a sequential wavelength dispersive spectrometer it is necessary to consider the counting error of the net response of the peak counting rate R_p and background counting rate R_b, since the analyte element is only responsible for $R_p - R_b$. Equation (9-8) must then be expanded [1] to include the background counting rate term:

$$s(R_p - R_b) = \frac{100}{\sqrt{t}} \frac{1}{\sqrt{R_p} - \sqrt{R_b}} \tag{9-9}$$

One of the conditions for equation (9-6) is that the total counting time t must be correctly proportioned between time spent counting on the peak, t_p, and time spent counting on the background, t_b:

$$\frac{t_p}{t_b} = \sqrt{\frac{R_p}{R_b}} \tag{9-10}$$

Several points are worth noting with reference to equation (9-9). Firstly, where the count time is limited—which is the case in most analyses—the net counting error is a minimum when $\sqrt{R_p} - \sqrt{R_b}$ is a maximum. This expression can therefore be used as a figure of merit for the setting of instrumental variables such as X-ray tube parameters, the choice of dispersion conditions, and so on. Secondly, it will be noted that as R_b becomes small relative to R_p, equation (9-9) approximates to equation (9-8). In other words, as the background becomes less significant relative to the peak, its effect on the net counting error becomes smaller. The point at which the peak to background ratio exceeds 10:1 is generally taken as that where background can be ignored completely. A third point to be noted is that as R_b approaches R_p the counting error becomes infinite, and this will be a major factor in determining the lowest concentration limit that can be detected. This introduces the concept of the lower limit of detection, which will be discussed in detail in section 9.5.

In the case of the energy dispersive spectrometer, the peak and background are recorded simultaneously and the question of division of time between them does not arise. A choice does, however, have to be made as to what portion of the complete recorded spectrum should be used for the measurement of peak and background [2]. The net counting error s_{net} associated with equation (9-7) is given by

$$s_{net} = \sqrt{P + B\left(1 + \frac{n_p}{n_b}\right)} \tag{9-11}$$

If each number of background channels is chosen to be one half of the total number of peak channels, equation (9-11) reduces to $s_{net} = \sqrt{P + 2B}$ or, expressed as a percentage of the peak, as:

$$s_{net} = \frac{100\sqrt{P + 2B}}{P} \tag{9-12}$$

Equation (9-12) is the formula normally quoted for the percentage standard deviation of the net peak intensity.

9.4   MATRIX EFFECTS

In the conversion of net line intensity to analyte concentration it is necessary to correct for any interelement interactions that my occur. In the case of homogeneous specimens these interelement effects fall into two broad categories: absorption effects and enhancement effects. Absorption effects include both secondary and primary absorption, and enhancement effects include direct enhancement, involving the analyte element and one enhancing element, as well as third element effects, involving additional element(s) beyond the analyte and enhancer. As was shown in section 1.4, the mass absorption coefficient is a number that represents the magnitude of the absorption of a certain element for a specific X-ray wavelength. Discontinuities in an absorption curve will occur at wavelengths (energies) corresponding to the binding energy values of the various atomic subshells of the absorbing element. Between these absorption edges, the mass absorption coefficient values vary roughly as $1/Z^3$, where Z is the atomic number of the absorber.

   Primary absorption occurs because all atoms of the specimen matrix absorb photons from the primary source. Since there is a competition for these primary photons by the atoms making up the specimen, the intensity–wavelength distribution of the photons available for the excitation of a given analyte element may be modified by other matrix elements. As an example of this, figure 9-2 shows three intensity–wavelength distributions of primary source photons being used for the excitation of matrix elements A and B. The upper diagram shows a hatched area under the smooth curve representing the distribution of the primary photons, before they strike the specimen. Also shown in the figure are the absorption curves and characteristic lines for elements A and B. Considering first the excitation of element A, as shown in the middle diagram, it will be seen that the portion of the primary continuum available for the excitation of A is indicated by the hatched area between the minimum wavelength of the continuum and the absorption edge of element A. Note that the hatched area is less intense than the original primary continuum because of the absorption effect of B. Consider now the excitation of B, as shown in the lower diagram. As in the case of A, the continuum available for the excitation of B lies between the minimum wavelength and the absorption edge of B, again decreased because of the absorption effect of A. This time, however, the distribution curve is not completely smooth and has a sharp discontinuity at the wavelength of the absorption edge of A. This is because the portion of the continuum to the short wavelength side of the absorption edge of A is strongly absorbed, and to the long wavelength side is weakly absorbed, following the shape of the absorption curve of A shown in the upper diagram. Thus the effect of primary absorption is to modify that portion of the spectrum most effective in the excitation of a given analyte. The degree of this modification is in turn dependent upon *all* matrix elements present.

Secondary absorption refers to the effect of the absorption of characteristic analyte radiation by the specimen matrix. As characteristic radiation passes out from the specimen in which it was generated, it will be absorbed by all matrix elements, in amounts related to the mass absorption coefficients of these elements. As an example, referring again to the upper diagram in figure 9-2, note where the absorption curves intersect the characteristic lines of elements A and B. As far as B is concerned, its characteristic line is intersected at rather low values by both A and B. On the other hand, looking at A, the absorption of B radiation is clearly far greater than that of A radiation. Thus, the effect is that element B strongly absorbs radiation from element A, and element A weakly absorbs its own radiation. This effect is called secondary absorption.

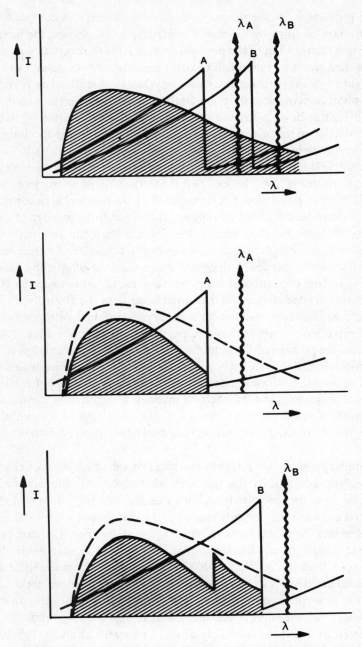

Figure 9-2. Fraction of total primary spectrum available for excitation.

The total absorption α of a specimen is dependent on both primary and secondary absorption. The total absorption by element j for an analyte wavelength i is given by the following relationship:

$$\alpha_i = \mu_i(\lambda) + A[\mu_i(\lambda_j)] \tag{9-13}$$

The factor A is a geometric constant equal to the ratio of the sines of the incident and takeoff angles of the spectrometer. This factor is needed to correct for the fact that the incident and emergent rays from the sample have different path lengths. The $\mu_i(\lambda)$ term in the equation refers to the primary radiation. Since most conventional X-ray spectrometers use a bremsstrahlung source, in practice $\mu_i(\lambda)$ is a range of wavelengths although in simple calculations it may be acceptable to use a single "equivalent" wavelength value [3], defined as one having the same excitation characteristics as the full continuum.

Looking now at the effects of enhancement in practice, there are a number of routes by which the analyte element can be excited, and these are illustrated in figure 9-3. Figure 9-3a shows the direct excitation of an analyte element i by the primary continuum P_1. There may also be excitation of the analyte by characteristic lines from the source, designated in figure 9-3b by P_2. Both the continuous and characteristic radiation from the source may be somewhat modified by Compton scattering, and the excitation by this modified source radiation is indicated in figure 9-3c by P_3. Enhancement effects (figure 9-3d) occur when a nonanalyte matrix element A emits a characteristic line that has an energy just in excess of the absorption edge of the analyte element. This means that the nonanalyte element in question is able to excite the analyte, giving characteristic photons over and above those produced by the primary continuum. This gives an increased, or enhanced, signal from the analyte. An example of this enhancement effect can be seen in the upper diagram of figure 9-2, where the characteristic line of from A lies to the short wavelength (high energy) side of B and can therefore excite B.

The last effect considered here is the so-called third element effect, shown in figure 9-3e. Here, a third element B is also excited by the source. Not only can B directly enhance i, but it can also enhance A, thus increasing the enhancing effect of A on λ_i. This last effect is called the *third element effect*. Table 9-2 shows some of the data published for the chrome–iron–nickel system [4] and illustrates the relative importance of the various excitation routes. Relating these data to figure 9-3, chromium is the element i, iron is the enhancer A, and nickel is the third element B. The data given are for

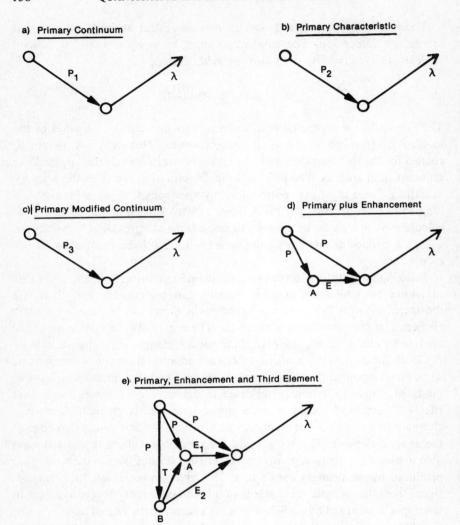

Figure 9-3. Processes leading to the excitation of characteristic photons.

three different alloy compositions: $25:25:50$, $10:40:50$ and $40:10:50$ nickel:iron:chromium. Note that in each case, roughly 87% of the actual measured chromium Kα radiation comes from direct excitation by the source. In the case of direct enhancement by iron, each percent of iron added increases the chromium radiation by about 0.26%. Direct excitation of chromium by nickel gives about a 0.21% increase in chromium radiation per

Table 9-2. Relative Importance of Primary Excitation, Enhancement and Third Element Effect in the Cr : Fe : Ni System[a]

| | Route[b] | Percentage of Counts for Chromium Kα | | |
		25:25:50	10:40:50	40:10:50
Primary	P	87.5	87.2	87.6
Secondary	E_1	6.7	10.6	2.6
	E_2	5.3	2.0	9.4
Third element	T	0.5	0.2	0.4

[a]Data taken from Shiraiwa, T., and Fujino, N. (1967), *Bull. Chem. Soc. Japan* **40**, 2289.
[b]Route indicated in figure 9-3.

percent nickel added. It will also be seen that the third element effect is much less important: here the chromium intensity is increase by only about 0.02% per percent of nickel added. From this one would predict that a "fourth element effect" would be negligible.

9.5 ANALYSIS OF LOW CONCENTRATIONS

The X-ray fluorescence method is particularly applicable to the qualitative and quantitative analysis of low concentrations of elements in a wide range of samples, as well as allowing the analysis of elements at higher concentrations in limited quantities of materials. The generally accepted definition for the lower limit of detection is that concentration equivalent to two standard deviations of the background counting rate (for a more detailed discussion of the derivation of the lower limit of detection for wavelength and energy dispersive spectrometers, see ref. 4, chapter 11). From the discussion in section 9.2 it follows that two standard deviations, $2s(N)$, of the total background counts N_b taken will be given by

$$2s(N) = 2\sqrt{N_b} = 2\sqrt{R_b t_b} \quad \text{(in counts)}$$

where t_b is the time spent counting on the background. To convert counts to count rate we divide by time; thus

$$2s(R) = \frac{2\sqrt{R_b t_b}}{t_b} = 2\sqrt{\frac{R_b}{t_b}} \quad \text{(in count rate)}$$

To convert count rate to concentration we divide by the sensitivity m:

$$2s(C) = \frac{2}{m}\sqrt{\frac{R_b}{t_b}} \quad \text{(in concentration)}$$

Since two measurements have to be made (peak and background), the error is increased by $\sqrt{2}$, and taking $2\sqrt{2} \sim 3$, we have the formula for the lower limit of detection, LLD:

$$\text{LLD} = \frac{3}{m}\sqrt{\frac{R_b}{t_b}} \quad (9\text{-}14)$$

Note that in equation (9-14) t_b represents one-half the total counting time. The detection limit expression for the energy dispersive spectrometer is similar to that for the wavelength dispersive system except that t_b now becomes the livetime of the energy dispersive spectrometer.

The sensitivity m of the X-ray fluorescence method is expressed in terms of the intensity of the measured wavelength per unit concentration, expressed in counts/sec per percent. Figure 9-4 shows the sensitivity of a wavelength dispersive spectrometer and indicates that the sensitivity varies by about four orders of magnitude over the measurable element range, when expressed in terms of rate of change in response per unit rate of change in concentration. For a fixed analysis time the detection limit is proportional to $m/\sqrt{R_b}$, and this is taken as a figure of merit for trace analysis. The value of m is determined mainly by the power loading of the source, the efficiency of the spectrometer for the appropriate wavelength and the fluorescent yield of the excited wavelength. The value of R_b is determined mainly by the scattering characteristics of the sample matrix and the intensity wavelength distribution of the excitation source.

It is important to note that not only does the sensitivity of the spectrometer vary significantly over the wavelength range of the spectrometer, but so too does the background counting rate. In general, the background varies by about two orders of magnitude over the range of the spectrometer. By inspection of equation (9-14) it will be seen that the detection limit will be best when the sensitivity is high and the background is low. Both the spectrometer sensitivity and the measured background vary with the average atomic number of the sample. While detection limits over most of the atomic number range lie in the low part per million range, the sensitivity of the X-ray spectrometer falls off quite dramatically towards the long wavelength limit,

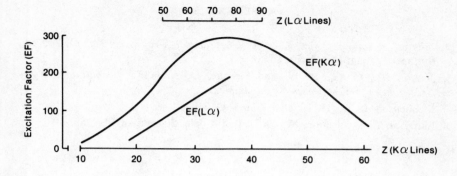

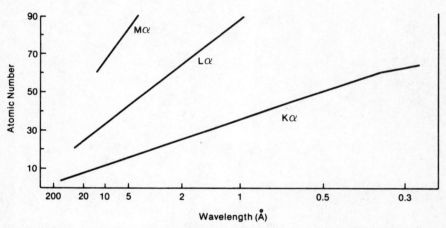

Figure 9-4. Sensitivity of the wavelength dispersive spectrometer as functions of atomic number and wavelength.

due mainly to low fluorescence yields and the increased influence of absorption. As a result, poorer detection limits are found at the long wavelength extreme of the spectrometer, which corresponds to the lower atomic numbers. Thus the detection limits for elements such as fluorine and sodium are at the levels of hundredths of one pecent rather than parts per million. The detection limits for the very low atomic number elements such as carbon ($Z = 6$) and oxygen ($Z = 7$) are, even worse—typically of the order of 3 to 5%.

REFERENCES

1. Jenkins, R., and de Vries, J. L. (1970). *Practical X-ray Spectrometry*, 2nd ed., Springer–Verlag: New York, Chapter 5.
2. Jenkins, R., Gould, R. W. and Gedcke, D. A. (1981). *Quantitative X-ray Spectrometry*, Dekker: New York, chapter 4.
3. Stephenson, D. A. (1971). *Anal. Chem.* **43**, 310.
4. Shiraiwa, T., and Fujino, N. (1967). Bull. Chem. Soc. Japan **40**, 2289.

CHAPTER

10

QUANTITATIVE PROCEDURES IN X-RAY FLUORESCENCE ANALYSIS

10.1 OVERVIEW OF QUANTITATIVE METHODS

In the X-ray analytical laboratory the quantitative method of analysis employed will be typically decided by a number of circumstances, of which probably the four most common are the complexity of the analytical problem, the time allowable, the computational facilities at the disposal of the analyst, and the number of standards available. It is convenient to break quantitative analytical methods down into two major categories: single element methods and multiple element methods, as shown in table 10-1. The simplest quantitative analysis situation to handle is the determination of a single element in a known matrix. A slightly more difficult case might be the determination of a single element where the matrix is unknown. As shown in the table, three basic methods are commonly employed in this situation: use of internal standards, use of a standard addition, or use of a scattered line from the X-ray source. The most complex case is the analysis of all, or most, of the elements in a sample about which little or nothing is known. In this case a full qualitative analysis will be required before any attempt is made to quantitate the matrix elements. Once the qualitative composition of the sample is known, again, one of three general techniques is typically applied: type standardization, an influence coefficient technique, or a fundamental parameter technique. Both the influence coefficient and fundamental parameter technique require a computer for their application.

In the X-ray fluorescence analysis of homogeneous specimens, the correlation between the characteristic line intensity of an analyte element and the concentration of that element is typically nonlinear over wide ranges of

143

Table 10-1. Quantitative Procedures Employed in X-Ray Fluorescence Analysis

Single Element Methods

Internal standardization
Standard addition
Use of scattered source radiation

Multiple Element Methods

Type standardization
Use of influence coefficients
Fundamental parameter techniques

concentration, due to interelement effects between the analyte element and other elements making up the specimen matrix. As was discussed in section 9.4, matrix effects are made up of various interferences and these can be rather complex. However, the situation can be greatly simplified in the case of homogeneous specimens, where severe enhancement effects are absent, and here, the slope of a calibration curve is inversely proportional to the total absorption α of the specimen for the analyte wavelength. In this instance the slope K of the calibration curve is taken as W/I, where I is the line intensity and W the weight fraction of the analyte element. Thus, the following relationship holds:

$$W = I\alpha K \qquad (10\text{-}1)$$

It follows from equation (10-1) that where a range of analyte concentrations must be covered, say from a low value W_ℓ to a high value W_h, with corresponding analyte intensities I_ℓ and I_h, the calibration curve will only be linear if the absorption α_ℓ for the low concentration matrix is equal to the absorption α_h for the high concentration matrix:

$$\frac{W_\ell}{W_h} = \frac{I_\ell}{I_h} \times \frac{\alpha_\ell}{\alpha_h} \times \frac{K_\ell}{K_h} \qquad (10\text{-}2)$$

The single element techniques, which will be discussed in detail later in this chapter, are all methods which *reduce* the influence of the absorption term in equation (10-2), generally by referring the intensity of the analyte wavelength to that of a similar wavelength, arising either from an added standard, or from a scattered line from the X-ray tube. In certain cases, limiting the concentration range of the analyte may allow the assumption to be made that

Table 10-2. Development of Computers Used in Controlled X-Ray Spectrometers,[a] **Showing Memory and Storage Available**

Approx. Year	Computer Memory (bytes)	Storage Medium	Programming Language
1965	4K	Paper tape	Assembler
1970	8K	Cassette	Assembler
1973	16K	Floppy disk ($\frac{1}{3}$ Mbyte)	High level/ assembler
1976	16K	Firm disk (5 Mbyte)	High level/ assembler
1980	32K	Firm disk (20 Mbyte)	High level
1985	640K	Firm disk (40 Mbyte)	High level

[a]Computer system and peripherals available for roughly 20% the cost of the total spectrometer system.

the values of K_t and K_h are nominally the same, i.e. that the calibration curve is essentially linear. This is assumption is applied in the traditional "type standardization" technique. Type standardization was widely employed in the 1960s and 1970s, but now that computers are generally available, it is usually considered more desirable to work with general purpose calibration schemes, which are applicable to a variety of matrix types over wide concentration ranges.

In 1955, Sherman [1] showed that it was possible to express the intensity concentration relationship in terms of independently determined "fundamental parameters." Unfortunately, fundamental type methods require a fair amount of computation, and in the mid 1950s suitable computational facilities were not available. Table 10-2 lists the typical computer systems which have been used in X-ray fluorescence analysis over the past 30 years or so. In the design of a computer controlled spectrometer, a rough guideline which is frequently employed is that the cost of the computer and its associated peripherals should not exceed 20% of the total cost of the spectrometer system.

As the cost, power and flexibility of the minicomputers increased, this newer technology was quickly taken advantage of by the spectrometer suppliers. It will be seen from the table that as the computational potential increased, there was a gradual changeover from the use of assembly code to high level languages, typically FORTRAN. Because of the limited computational facilities available in the early 1960s, and because of the clear need for some degree of fast mathematical correction of matrix effects, a number of so-called "empirical correction" techniques were developed which required far less computation and were therefore usable by the computers available at the time.

In principle, an empirical correction procedure can be described as the correction of an analyte element intensity for the influence of an interfering element(s), using as the correction term the product of the intensity from the interfering element line and a constant factor [2, 3]. This constant factor is generally referred to as an *influence coefficient*, since it is assumed to represent the influence of the interfering element on the analyte. Commonly employed influence coefficient methods may use either the intensity or the concentration of the interfering element as the correction term. These methods are referred to as *intensity correction* and *concentration correction* methods respectively. Intensity correction models give a series of linear equations which do not require much computation, but they are generally not applicable to wide ranges of analyte concentration. Various versions of the intensity correction models [4] found initial application in the analysis of nonferrous metals, where correction constants were applied as lookup table. Later versions [5] were supplied on commercially available computer controlled spectrometers and were used for a wider range of application [6]. The Lachance–Traill model [7] is a concentration model, which in effect requires the solving of a series of simultaneous equations, by regression analysis or matrix inversion techniques. This approach is more rigorous than the intensity models, and became rather popular in the early 1970s as suitable low cost minicomputers became available.

10.2   MEASUREMENT OF PURE INTENSITIES

The intensity of an analyte line is subject not only to the influence of other matrix elements, but also to random and systematic errors due to the spectrometer and counting procedure employed. Provided that a sufficient number of counts is taken, and provided that the spectrometer source is adequately calibrated, *random errors* from these sources are generally insignificant relative to other errors. *Systematic errors* from these sources are, however, by no means insignificant, and effects such as counting dead time, background, and line overlap can all contribute to the total experimental error in the measured intensity. A problem may arise in that incorrect conclusions about potential matrix effects may be drawn from data which are subject to instrumental systematic errors. As an example, under a certain set of experimental conditions, a series of binary alloys may give a calibration curve of decreasing slope. The conclusion may be drawn that radiation from the analyte element was enhanced by the other matrix element, where in point of fact the problem could have also been due to dead time loss in the counting circuitry because the count rate was too high. If a correction were applied for an enhancement effect, the procedure would break down if the count rates were changed, for example, by varying the source conditions. Problems of this type can be particularly troublesome in the application of influence correction methods. Unless the instrument dependent errors are completely separated from the matrix dependent terms, the instrument effects will tend to become associated with the influence correction terms. In practice, this may not be completely disastrous for a specific spectrometer calibrated for a particular application, since the method will probably work, provided that the experimental conditions do not change. However, one major consequence is that it will probably be impossible to "transport" a set of correction constants from one spectrometer to another. Mainly for this reason, it is common practice today to attempt to obtain intensities as free from systematic instrumental error as possible. Such intensities are referred to as "pure" intensities.

10.3   SINGLE ELEMENT METHODS

10.3.1   Use of Internal Standards

One of the most useful techniques for the determination of a single analyte element in a known or unknown matrix is to use an internal standard. The technique is one of the oldest methods of quantitative analysis [8] and is based on the addition of a known concentration of an element which gives a wavelength similar to that of the analyte wavelength. Referring to equation (10-2), if ℓ is replaced by the analyte element x, and h by the internal standard s, provided that the internal standard is selected such that the absorptions of the matrix for x and ℓ are about the same (i.e. $\alpha_\ell = \alpha_s$), then the following relationship is true:

$$\frac{W_x}{W_s} = K_i \frac{I_x}{I_s} \qquad (10\text{-}3)$$

where I_x and I_s are the measured intensities of analyte and internal standard wavelengths, and W_s the known added weight of the internal standard. The value of K_i is equal to K_x/K_s, that is, the ratio of the sensitivities of the spectrometer for the analyte and internal standard elements respectively. These data can be obtained by separate experiments.

Internal standards are best suited to the measurements of analyte concentrations below about 10%. This limit arises because it is generally advisable to add the internal standard element at about the same concentration level as that of the analyte. When more than 10% of the internal standard is added, it may significantly change the specimen matrix and introduce errors into the determination. Care must also be taken to ensure that the particle sizes of specimen and internal standard are about the same, and that the two components are adequately mixed. If these conditions are not met, the use of the internal standard may well correct for the absorption of the matrix, but it may also introduce problems of heterogeneity, as were discussed in section 7.2.

10.3.2 Standard Addition Methods

It sometimes happens that an appropriate internal standard cannot be found. In that case it may be possible to use the analyte itself as an internal standard. This method is a special case of standard addition, and it is generally referred to as *spiking*. Again referring to equation (10-2), if 1 refers to the unknown specimen x, the unknown concentration W_x will give a measured intensity of I_x. On deliberately adding an additional amount a of the analyte element, the new intensity of x will equal I_{x+a}. Thus in the equation, W_h is replaced by $W_x + W_a$, and I_1 becomes I_{x+a}. Since the absorption coefficient is essentially the same for both matrices, before and after addition, and since the slope constants are also identical, it follows that

$$\frac{W_x}{W_x + W_a} = \frac{I_x}{I_{x+s}} \tag{10-4}$$

Since I_x and I_{x+a} are both measured, and W_a is known, a value for W_x can easily be obtained.

The technique of standard addition is especially useful for the determination of analyte concentrations below about 5%. It is also useful in the preparation of secondary standards. However, care must be taken in the application of the method whenever the relationship between analyte concentration and analyte intensity is nonlinear. It will be apparent from inspection of equation (10-4) that a linear relationship is assumed between concentration and intensity, and if this is not the case, errors will occur. Since one may not be sure how linear the concentration–intensity relationship is in a given situation, it is common practice to repeat the standard addition step at least twice, so that a minimum of three data sets (zero concentration plus the two additions) are available to work with.

10.3.3   Use of Scattered Tube Lines

It was been shown in the previous chapter that the background intensity in a given measurement is always present, due mainly to scattering, and under certain circumstances it may be possible to pick out a wavelength region in the background and use it as an "internal standard" [9]. It was also stated in the previous chapter that the scattered background B varies approximately as $1/Z^2$; thus

$$BZ^2 = K_1 \qquad (10\text{-}5)$$

Since the mass absorption coefficient μ varies as the cube of the atomic number Z, that is,

$$\mu = K_3 Z^3 \qquad (10\text{-}6)$$

and since the slope m of a calibration curve is inversely proportional to the absorption of the specimen matrix, we have

$$m = \frac{I}{C} = \frac{K_2}{\mu} \qquad (10\text{-}7)$$

It follows that by combining equations (10-6) and (10-7) an expression can be obtained for concentration (C) in terms of the measured peak to background ratio, i.e.,

$$I = \frac{C}{Z^3} \frac{K_2}{K_3} \qquad (10\text{-}8)$$

It will be seen from equation (10-8) that there is a very strong dependence of the slope of the calibration curve on the average atomic number of the specimen—in fact, as the third power of Z. Combining equations (10-5) and (10-8) gives

$$\frac{I}{B} = \frac{C}{Z}\left[\frac{K_2}{K_1 + K_3}\right] \tag{10-9}$$

Note from equation (10-9) that the peak to background ratio is far less dependent on the average atomic number (although not independent). Thus, while use of a scattered tube line as a standard may not completely ovecome a matrix effect, it will generally reduce it significantly.

As an example, the data listed in table 10-3 were obtained from a series of standards which were made to evaluate the use of scattered background as an internal standard for uranium ore analysis [10]. A series of matrices were chosen, and 100 ppm of uranium was added to each. The experimentally observed sensitivities in counts/sec per percent for uranium is listed in the second column of the table; it will be seen that, because of the wide range in matrix absorption coefficients, the sensitivity varies by about a factor of 6. There is a similarly wide variation in the background count rate, this time due to the large difference in the average atomic number Z. If one simply attempted to use the first sample (that is, the SiO_2) as the standard, and ratioed the observed count rate on each of the other four matrices to it to obtain a measure of the concentration, values widely varying from the expected value of 100 ppm would be obtained. These results are listed in table 10-3 as "uncorrected." However, it follows from equation (10-9) that if the values of K_1, K_2 and K_3 really are constant, and the atomic number term Z is ignored, the concentration should be proportional to the peak to background ratio:

$$C = K_4 \frac{I}{B} \tag{10-10}$$

Again using the SiO_2 as a reference standard, a new set of concentrations are obtained, listed as "corrected" in the table. While these corrected data are not exactly equal to 100 ppm, they are, in most cases, probably good enough for routine work, especially when one considers that the standards used represent the extremes of likely ore composition.

Table 10-3. Use of Scattered Background as an Internal Standard in the Analysis of Uranium Ores[a]

Matrix	Sensitivity [(counts/sec)/%]	R_b	R_p/R_b	Z	M.A.C.[b]	Conc. (ppm) Uncorrected	Conc. (ppm) Corrected
SiO_2	5.1	390	2.31	10.5	7.9	100[c]	100[c]
SiO_2–Fe_2O_3	1.9	150	2.15	17	30.6	268	93
Fe_3O_4	0.9	98	1.87	21	55	567	81
$CaCO_3$	2.4	220	2.01	13	16.9	213	91
$KAlSi_3O_6$	3.8	304	2.23	12	12.2	134	97

[a] Data taken from James, G. W. (1977). *Anal. Chem.* **49**, 967.
[b] M.A.C. = matrix absorption coefficient.
[c] Used as comparison reference.

10.4 MULTIPLE ELEMENT METHODS

10.4.1 Type Standardization

As has been previously stated, provided that the total specimen absorption does not very significantly over a range of analyte concentrations, and provided that enhancement effects are absent and that the specimen is homogeneous, a linear relationship will be obtained between analyte concentration and measured characteristic line intensity. Where these provisos are met, type standardization techniques can be employed. It will also be clear from previous discussion that by limiting the range of analyte concentration to be covered in a given calibration procedure, the range in absorption can also be reduced. Type standardization is probably the oldest of the quantitative analytical methods employed, and the method is usually evaluated by taking data from a well-characterized set of standards, and, by inspection, establishing whether a linear relationship is indeed observed. Where this is not the case, the analyte concentration range may be further restricted. The analyst of today is fortunate in that many hundreds of good reference standards are commercially available [11]. While the type standardization method is not without its pitfalls [12], it is nevertheless extremely useful, especially for quality control applications where a finished product is being compared with a desired product.

Special reference standards may be made up for particular purposes, and these may serve the dual purpose of instrument calibration as well as establishing working curves for analysis. As an example, two thin glass film standard reference materials specially designed for calibration of X-ray spectrometers are available from the National Bureau of Standards in Washington as Standard Reference Materials (SRM) 1832 and 1833 [13]. They consists of a silica-base film deposited by focused ion beam coating onto a polycarbonate substrate. SRM 1832 contains aluminum, silicon, calcium, vanadium, manganese, cobalt and copper; SRM 1833 contains silicon, potassium, titanium, iron, zinc and rhodium. The standards are especially useful for the analysis of particulate matter [14].

10.4.2 Influence Correction Methods

Lachance [15] has suggested dividing influence coefficient correction procedures into three basic types: fundamental, derived and regression. Fundamental models [16, 17] are those which require starting with concentrations

Linear Model:
$$W_i/R_i = K_i$$

Lachance-Traill (1966):
$$W_i/R_i = K_i + \sum_j a_{ij}W_j$$

Claisse-Quintin (1967):
$$W_i/R_i = K_i + \sum_j a_{ij}W_j + \sum_j \gamma_{ij}W_j^2$$

Rasberry-Heinrich (1974):
$$W_i/R_i = K_i + \sum_j a_{ij}W_j + \sum_{k \neq j} \beta_{ik}(W_k/1 + W_i)$$

Lachance-Claisse (1980):
$$W_i/R_i = 1 + \sum_j a_{ij}W_j + \sum_j \sum_{k>j} a_{ijk}W_j\, W_k$$

Figure 10.1. Commonly employed influence correction models.

and then calculating the intensities. Derived models are those which are based on some simplification of a fundamental method [7, 18] but which still allow concentrations to be calculated from intensities. Regression models are those which are semiempirical in nature [5, 19] and which allow the determination of influence coefficients by regression analysis of data sets obtained from standards (e.g. ref. 20). All regression models have essentially the same form and consist of a weight fraction term W (or concentration C), an intensity (or intensity ratio) term I, an instrument dependent term which essentially defines the sensitivity of the spectrometer for the analyte in question, and a correction term which corrects the instrument sensitivity for the effect of the matrix. The general form is as follows:

$$W = I\langle\text{instrument}\rangle\left(1 + \sum\{\text{model}\}\right) \qquad (10\text{-}11)$$

The different models vary only in the form of the correction term. Figure 10-1 shows several of the more important commonly employed influence coefficient methods. All of these models are concentration correction models in which the product of the influence coefficient and the *concentration* of the interfering element is used to correct the slope K_i of the analyte calibration curve.

The Lachance–Trail model [7] was the first of the concentration correction models to be published. Some years after it appeared, Heinrich and his co-workers at the National Bureau of Standards suggested an extension [19] in which absorbing and enhancing elements are separated as α and β terms. These authors suggested that the enhancing effect cannot be adequately described by the same hyperbolic function as the absorbing effect.

A thorough study [16] of Lachance–Trail coefficients based on theoretically calculated fluorescence intensities shows that all binary coefficients vary systematically with composition. Both the Claisse–Quintin [18] and Lachance–Claisse [21] models use higher order terms to correct for so-called "crossed effects", which include enhancement and third element effects. These models are generally more suited for very wide concentration range analysis.

In all of these methods one of three basic approaches is used to determine the values of the influence coefficients, following the initial measurement of intensities using a series of well-characterized standards. The first approach is to use multiple regression analysis techniques to give the best fit for slope, background and influence coefficient terms. Alternatively, the same data set can be used to graphically determine individual influence coefficients. As an example, in the case of the Lachance–Trail equation, for a binary mixture a–b the expression for the determination of a would be

$$W_a/I_a = 1 + \alpha_{ab}W_b \qquad (10\text{-}12)$$

By plotting data from a range of analyzed standards in terms of W_a/R_a as a function of W_b, a straight line should be observed with a slope of α_{ab} and an intercept of unity. This approach is especially useful for visualizing the form of the influence coefficient correction [22]. Thirdly, the influence coefficient can be calculated using a fundamental type equation based on physical constants (see section 10.4.3).

Several useful extensions to the basic models have been proposed. For example, a FORTRAN program (NBSGSC) for quantitative X-ray spectrometry has been described by Tao et al. [23]. This program is able to handle pressed or fused samples; the ability to process data from fused samples comes because in mathematical correction processes using influence coefficients, since the condition $\sum w = 1$ gives $n + 1$ equations, one term can be dropped (one usually chooses α_{ii}). One can make use of this to avoid having to analyze all of the matrix, e.g. α_x for ignition loss, etc. This will, however, change the other α's.

The major advantage to be gained by the use of influence coefficient methods is that a wide range of concentration ranges can be covered using a fairly inexpensive computer for the calculations. A major disadvantage is that a large number of well-analyzed standards may be required for the initial determination of the coefficients. However, where adequate precautions have been taken to ensure correct separation of instrument and matrix dependent terms, the correction constants are transportable from one spectrometer to another and, in principle, need only be determined once.

While influence coefficient methods have gained a good deal of popularity, workers still sometimes encounter difficulties in their application, and there are a number of typical reasons why they may fail [24]. The most common of these are:

1. Failure to adequately separate instrumental and matrix dependent effects.

2. Poor judgement on the part of the analyst as to whether or not a correction term should really be included.

3. Poor technique on the part of the analyst in the determination of the influence coefficients.

4. Poor quality and/or range of calibration standards.

5. Inadequacy of the regression analysis program used in the determination of the coefficients.

6. Application of the technique in cases where the specimens are insufficiently homogeneous.

10.4.3 Fundamental Methods

Since the early work of Sherman [1] there has been a growing interest in the provision of an intensity concentration algorithm which would allow the calculation of the concentration values without recourse to the use of standards. Sherman's work was improved upon first by Shiraiwa and Fujino [25], and later by Criss and Birks [26, 27] with their program NRLXRF. The latter group also solved the problem of describing the intensity distribution from the X-ray tube [28]. The problem for the average analyst in the late 1960's and early 1970's, however, remained that of finding sufficient computational power to apply these methods. In the early 1970s, de Jongh suggested an elegant solution [29] in which he proposed the use of a large mainframe computer for the calculation of the influence coefficients, followed by the use of a small minicomputer for their actual application using a concentration correction influence model. Large tables of correction constants determined by this method are now available [30]. One of the problem areas remains that of adequately describing the intensity distribution from the X-ray tube. Gilfrich and Birks demonstrated an experimental approach to this problem by measuring the spectral distribution from the tube in an independent experiment [28]. More recently, this work has been extened to calculate spectral distributions using data obtained from the electron microprobe [31].

While software packages are available for fundamental calculations using data obtained with the energy dispersive system, one major drawback remains in their application systems, which use a modified primary excitation spectrum. Most fundamental quantitative approaches in use today employ measured or calculated continuous radiation functions in the calculation of the primary absorption effect. Where sharp discontinuities or "breaks" in this primary spectrum occur—as in the case of the energy dispersive system—the calculation becomes very complicated. This is probably the reason why many energy dispersive based quantitative procedures not covered by type standardization tend to favor simpler, though less accurate, methods based on scattered tube lines.

REFERENCES

1. Sherman, J. (1955). *Spectrochim. Acta* **7**, 283.
2. Beattie, M. J., and Brissey, R. M. (1954). *Anal. Chem.* **26**, 980.
3. Gillam, E., and Heal, H. T. (1952). *J. Appl. Phys.* 353.
4. Lucas-Tooth, H. J., and Price, B. J. (1961). *Metallurgia*, **54**, 149.
5. Lucas-Tooth, H. J., and Pyne, C. (1964). *Adv. X-Ray Anal.* **7**, 523.
6. Jenkins, R., de Klerck, J., and van Gelder, S. (1970). Philips S.&A.E. Bulletin FS-28, Philips: Almelo, Netherlands.
7. Lachance, G. R., and Traill, R. J. (1966). *Can. Spectrosc.* **11**, 43.
8. L. S. Birks (1959). *X-ray Spectrochemical Analysis*, Interscience Chemical Analysis Series, Volume 11, Interscience, New York, p. 65.
9. Anderman, G., and Kemp, J. W. (1958). *Anal. Chem.* **30**, 1306.
10. James, G. W. (1977). *Anal. Chem.* **49**, 967.
11. Lists of commercially available calibration standards for use in X-ray Spectrometry were published in *The International Journal of X-Ray Spectrometry*, Volumes 6, 7, 8 (1977–1977), Wiley/Heyden: London.

12. R. Jenkins, R. W. Gould, and D. A. Gedcke (1981). *Quantitative X-Ray Spectrometry*, Dekker: New York, p. 450.

13. Pella, P. A., Newbury, D. E., Steel, E. B., and Blackburn, D. H. (1986). *Anal. Chem.* **56**, 1133.

14. Hayaska, T., Shibata, Y., Inone, Y., Hayashi, H., and Kurosawa, Y. (1985). *Kawasaki-shi Kogai Kenkyusho Nenpo* **12**, 13.

15. Lachance, G. R., private communication.

16. Tertian, R. (1986). *X-Ray Spectrom.* **15**, 177.

17. Rousseau, R. (1984). *X-Ray Spectrom.* **13**, 115.

18. Claisse, F., and Quintin, M. (1967). *Can. Spectrosc.* **12**, 159.

19. Rasberry, S. D., and Heinrich, K. F. J. (1974). *Anal. Chem.* **46**, 81.

20. Schreiner, W. N., and Jenkins, R. (1979). *X-Ray Spectrom.* **8**, 31.

21. Lachance, G. R., and Claisse, F. (1980). *Adv. X-Ray Anal.* **23**, 87.

22. G. Lachance (1984). *Introduction to Alpha Coefficients*, Corporation Scientifique Claisse: Sainte-Foy, Quebec.

23. Tao, G. Y., Pella, P. A., and Rousseau, R. M. (1985). *Gov. Rep. Announce. Index* (*U.S.*) **85**(17), Abstr. No. 537, 742.

24. Jenkins, R. (1979). *Adv. X-Ray Anal.* **22**, 281.

25. Shiraiwa, T., and Fujino, N. (1967). *Bull. Chem. Soc. Japan* **40**, 2289.

26. Criss, J. W. and Birks, L. S. (1968). *Anal. Chem.* **40**, 1080.

27. Criss, J. W. (1980). *Adv. X-Ray Anal.* **23**, 93.

28. Gilfrich, J. V., and Birks, L. S. (1968). *Anal. Chem.* **40**, 1077.

29. de Jongh, W. K. (1973). *X-Ray Spectrom.* **2**, 151.

30. *Alpha Tables*, available from N. V. Philips: Almelo, the Netherlands.

31. Pella, P. A., Feng, L. Y., and Small, J. A. (1985). *X-Ray Spectrom.* **14**, 125.

APPLICATION OF X-RAY ANALYTICAL METHODS

11.1 COMPARISON OF X-RAY FLUORESCENCE WITH OTHER INSTRUMENTAL METHODS

Since the early growth of instrumental methods of analysis in the 1950's and 1960s, there has always been a conceived need to develop a general purpose method of analysis. By this is meant a single technique that can be applied to the analysis of any combination of elements in any matrix, yielding all of the information the analyst would like to have, and giving the ultimate in precision and accuracy. During the past 30 years or so a number of methods have been developed which, though they initially seemed to satisfy all of the needs of the analyst, still fell short in certain areas. In point of fact, almost never is an analytical technique completely obsoleted by a newer one. Like most other branches of science, development in analytical instrumentation is generally evolutionary rather than revolutionary. While there may be significant improvements in (for example) detectors or sources, which bring about dramatic improvement, in the long run, more often than not, techniques end up complementing one another rather than replacing one another. All of this can create a problem for the analyst, who has to choose, from the vast array of techniques available, the one that is best for his or her application.

The techniques currently available for the quantitative determination of elements include [1, 2] atomic absorption, atomic fluorescence, flame emission spectrometry, mass spectrometry, X-ray fluorescence, electrochemistry, emission spectrometry, and nuclear and radiochemical analysis. Nearly all of these techniques are based on the study of either radiation emission or radiation absorption by a sample. Techniques that involve study of radiation emission include X-ray fluorescence, in both the wavelength and the energy dispersive modes; and ultraviolet emission. The fundamental differences between the X-ray and ultraviolet techniques arise mainly from the wavelength of the excited radiation. In X-ray methods the radiation characteristic of the atom arises from transitions from atomic orbitals from the inner shells of the atom. This means that the characteristic wavelengths are largely free from bonding effects and spectra are rather simple. On the other hand, longer wavelength X-rays are difficult to disperse and even more difficult to detect quantitatively, and there can be significant problems in the analysis of elements of atomic number less than 10. Ultraviolet emission arises from electron transitions to outer atomic orbitals, and while the spectra are more complex, it is much easier to disperse, focus and detect ultraviolet and visible light than it is X-rays.

The techniques that involve absorption of discrete wavelengths include flame and furnace atomic absorption spectroscopy and inductively coupled plasma spectroscopy. The key differences between these methods stem from the temperature at which the sample can be held during the absorption process. In general, a technique that offers better dissociation of the element gives absorption data that are much freer from interferences than one where the element is still tightly bound to its neighbors. Better dissociation generally means higher temperatures; thus the plasma sources clearly hold advantages over flames. Sometimes, advantage may be gained by combining two techniques, for example, X-ray excited ultraviolet emission, gas chromatography with mass spectroscopy (GC/MS), and, more recently, inductively couple plasma spectroscopy with mass spectroscopy (ICP/MS).

The ultraviolet emission spectrometer based on arc and spark sources has been the workhorse of the instrumental elemental analysis laboratory for more than four decades. The wide range of application of this method, in terms of range of elements analyzable, sensitivity, speed and accuracy, have made this the method of choice in thousands of laboratories all over the world. A major drawback, however, has always been the problem of the introduction of the sample into the excitation system. This problem has spawned a remarkable range of sample pretreatments and introductions, including ultrasonic nebulization, microliter flow injectors, wire loop microfurnaces, furnace atomizers, electrothermal vaporization, and a whole host of others.

Flame atomic absorption (AA) was developed in the early 1960s and today is one of the most useful of the instrumental methods for elemental analysis. There are few interferences, and what there are can be fairly easily controlled.

Standardization is usually simple. Elements that are not easily determined by the flame AA method include the more refractory elements that are only partly dissociated in the flame, such as boron, vanadium, tantalum and tungsten, and elements that have their resonance lines in the far ultraviolet, including phosphorus, sulfur and the halogens. Better dissociation can be achieved by use of plasmas rather than flames, and for such elements the inductively coupled plasma (ICP) method is by far the most commonly used. An additional advantage is that detection limits are better by about a factor of three for ICP than for flame AA. The third commonly used AA method is the graphite furnace technique. On a relative basis, furnace AA is has better detection limits, by one or two orders of magnitude, than flame AA or ICP. On an absolute mass basis, it is generally better than either of the other two methods, by as much as orders of magnitude. While a major drawback in graphite furnace AA has been the complex matrix effects, these problems are now better understood, and methodologies are being developed to control or overcome them. Unfortunately, furnace determinations are slow and are typically single element. Analysis times typically run to several minutes.

While not many detailed studies have been reported comparing these various techniques with X-ray fluorescence, in general the sensitivities and detection limits are of the same order. As an example, in a comparative study between energy dispersive X-ray spectrometry and AA spectrometry for the determination of metals in suspended particulates, detection limits for manganese, iron, nickel, copper, zinc and lead were found to be 0.01–0.05 $\mu g/cm^2$. The concentration data were about 35% lower than the equivalent X-ray data, and this was attributed to the incomplete solution of solid matter before analysis by AA spectrometry [3]. Similarly, in a comparative study between X-ray spectrometry and ICP atomic emission spectrometry, 5–10% precision has been reported for both methods, for trace element concentrations down to 1–10 $\mu g/g$. Samples used in this study included soils and grasses [4].

The traditional use of the mass spectrometer has involved analysis in the gaseous phase, and has been mostly restricted to volatile organic liquids. However, a dramatic change has taken place in the last several years which now allows study of the chemistry and analysis of biological substances in condensed polar phases. A whole new field has grown up with the advent of plasma and laser desorption followed by secondary ion bombardment of the liquid surface layer with, for example, energetic primary ions in the technique of Laser Secondary Ion Mass Spectrometry (LSIMS). While the sensitivity of these so-called "soft ionization techniques" cannot compete yet with techniques such as GC/MS, indications are that new developments in ion optical systems may eventually bring the sensitivity of the technique down from the existing nanomole range to the picomole range.

While features such as accuracy, sensitivity, and cost are probably paramount in the selection of a technique, one factor of growing importance is that of automation. A technique that is easily automated will find far greater acceptance than one that is not. Perhaps the best example of this is found in atomic absorption. Flame atomic absorption units are not easily automated for the analysis of simultaneous elements in a cost effective manner, mainly because of the need to change light sources. As a result, almost all AA instruments are designed for the analysis of single elements. This is, of course, not to say that flame AA absolutely cannot be automated. For example, there are units available that can analyze six elements in 50 samples in 30–40 minutes. However, this might not be the method of choice here when one considers that with the modern ICP system, which does not need to change sources between different elements, one can analyze perhaps 40 to 50 elements per minute. In addition to the question of the ease and cost of automation, another constraint to be considered is the form and size of the sample needed for the analysis. Some techniques are ideally, but not exclusively, suited to the analysis of liquids and solutions: flame AA and ICP are good examples. Other, such as X-ray fluorescence, are ideally, but not exclusively, suited to the analysis of solid samples.

11.2   APPLICATIONS OF X-RAY SPECTROMETRY

The great flexibility and range of the various types of X-ray fluorescence spectrometer, coupled with their high sensitivity and good inherent precision, make them ideal for quantitative analysis. Single channel wavelength dispersive spectrometers are typically employed for both routine and nonroutine analysis of a wide range of products, including ferrous and nonferrous alloys, oils, slags and sinters, ores and minerals, thin films, and so on. These systems are very flexible, but relative to the multichannel spectrometers are somewhat slow. The multichannel wavelength dispersive instruments are used almost exclusively for routine, high throughput analyses where the great need is for fast accurate analysis, but where flexibility is of little importance. Energy dispersive spectrometers have the great advantage of being able to display information on all elements at the same time. They lack somewhat in resolution over the wavelength dispersive systems, but also find great application in quality control, trouble shooting, and so on. They have been particularly effective in scrap alloy sorting, in forensic science and in the provision of elemental data to supplement X-ray powder diffraction data.

An area where X-ray techniques are finding increasing application is in the analysis of pollutants. For many years the sensitivity of the fluorescence method was barely sufficient for the direct measurement of contaminant levels at the ppm level, and preconcentration methods had to be employed. More recent developments have allowed sensitivities that are quite sufficient for the sample masses typically encountered in the analysis of trace metals in air and water samples, and X-ray fluorescence methods find increasing application in this area [5]. A good example of the wide applicability of the X-ray fluorescence method is in the paper industry, where it has been used for a broad range of applications, including the analysis of lime and lime related compounds; calcite; dolomites; silicates; silica glass sand; gypsum; MnO_2 and iron ores; wood treated with pentachlorophenol and ammonical copper arsenate; sulfur, chlorine, bromine and titanium in paper, pulp, pitch and asphalt; and evaluation of paper printability and ink holdout [6].

Because of the nondestructiveness of X-ray analysis the technique has found special usefulness in archeometry. Examples include the examination of the chemical composition of medieval durable blue soda glass from York Minster in England. This study showed three distinct compositional groups, indicating in turn three probable sources of the glass [7]. Energy dispersive spectrometer studies of early American pottery have indicated details of manufacturing location and time period [8]. A study of old porcelain revealed that the chemical composition of the ceramic body, the glaze and the various colors, could be used as a means of authenticating eighteenth and nineteenth century pieces [9]. Similar energy dispersive methods have been used for authenticating fine Chinese porcelain from the Ming (1348–1644) and Qing (1644–1911) dynasties [10]. It has been reported that all Chinese porcelains from Kangxi (1662–1722) up to the time of World War II have barium contents in the range 100–130 ppm. After this time the barium content varied from 60 to 700 ppm, with only a few pieces in the 100–130 ppm range. It is thus possible to rapidly identify most modern fake reproductions [11].

There are certain areas of elemental analysis that are still unique to X-ray fluorescence. As an example, nondispersive X-ray fluorescence has been used for the analysis of technetium in solutions by use of the Tc Kα line. At this time there are no good chemical methods for the analysis of technetium [12].

Direct X-ray fluorescence methods are also being used for the study of the short lived, superheavy elements. Another unique application of the fluorescence technique is in the direct analysis of biological samples. As an example, radioisotope X-ray fluorescence has been used for the analysis of biological samples such as those provided by biopsy of the human aorta or tissue sections. A collimated [109]Cd source of 10–20 mCi strength gave a sensitivity of about 10^{15} atoms and is being used for tumor growth studies [13] and studies involving atherosclerosis [14].

In addition to the many hundreds of papers that are published annually describing specific applications of the technique [15,16], and a journal specifically devoted to the fluorescence method [17], several texts have been devoted to specific areas of application. These include geology [18], medicine [19], and metallurgy and art [20].

11.3 COMBINED DIFFRACTION AND FLUORESCENCE TECHNIQUES

Like all analytical techniques, both X-ray powder diffraction and X-ray fluorescence have their advantages and disadvantages. Table 11-1 indicates the more important features of the two techniques. In terms of the range of application, X-ray fluorescence analysis allows the quantitation of all elements in the periodic table from fluorine (atomic number 9) upwards. Accuracies of a few tenths of one percent are possible, and elements are detectable in most cases to the low ppm level. X-ray powder diffractometry is applicable to any ordered (crystalline) material and, although much less accurate or sensitive than the fluorescence method, is almost unique in its ability to differentiate phases. The techniques differ widely in terms of their speed of analysis. The modern multichannel wavelength dispersive spectrometer is able to produce data from 20–30 elements in less than one minute. An energy dispersive spectrometer, or even a relatively simple wavelength dispersive spectrometer, is able to perform a full qualitative analysis on an unknown sample in less than 30 minutes. The diffraction technique, on the other hand, can take as much as an order of magnitude longer than this. Even with the most sophisticated computer controlled powder diffractometers available today, diffraction experiments are invariably very time consuming. A similar difference also exists in the sensitivities of the two methods. Whereas the fluorescence method is able to measure signals from as little as one part per million of a given element, the diffraction technique is often hard put to measure one percent.

Table 11-1. Features of X-Ray Spectrometry and Diffraction

Feature	Spectrometry	Diffractometry
Range	$Z > 5$	All ordered materials
Speed	sec–min	min–hr
Precision (%)	0.1%	0.25%
Accuracy (%)	0.1–1.0	0.5–5
Sensitivity	Low ppm	0.1–2
Cost ($)	50,000–250,000	25,000–125,000

In terms of precision and accuracy, again the spectrometry method out-performs the diffraction technique, perhaps by as much as an order of magnitude. The diffraction technique is at best a rather insensitive, slow technique, giving somewhat poor quantitative accuracy. On the other hand, it will be realized that the information given by the X-ray diffraction method is unique: no other technique is able to provide such data. This is not true of the X-ray fluorescence method, since, as was discussed in section 11.1, there are many other techniques available to the analytical chemist today for the quantitation of elements. The fluorescence and diffraction techniques are to a large extent complementary, since one allows accurate quantitation of elements to be made and the other allows qualitative and semiquantitative estimations to be made of the way in which the matrix elements are combined to make up the phases in the specimen.

As was discussed in chapter 8, both the simultaneous wavelength dispersive spectrometer and the energy dispersive spectrometer lend them-selves admirably to the qualitative and semiquantitative analysis of materials, because the characteristic X-ray spectra are so simple the actual process of allocating atomic numbers to the emission lines is a simple process and the chance of making a gross error is rather small. By comparison, the procedures for the qualitative analysis of multiphase materials with the X-ray powder diffractometer is a much more complex business. There are only one hundred or so elements, and within the range of the conventional spec-trometer each element gives, on an average, only half a dozen lines. In diffraction, on the other hand, there are perhaps as many as several million possible compounds, each of which can give on an average fifty or so lines. However, a combination of spectrometric and diffraction data can often be used to greatly simplify what would otherwise be a complex and time consuming analysis. By judicious use of both fluorescence and diffraction data, a "material balance" can often be established. The fluorescence data are first used to establish the higher concentration elements, and then the diffraction data are used to suggest how these elements might be combined in the analyzed specimen.

11.4   ON STREAM ANALYSIS

The area of process analytical chemistry deals with the problems of the provision of qualitative and quantitative information about a given chemical process [21]. Such data may be collected off line, at line or on line, and X-ray fluorescence is applicable in all three of these cases. This section deals mainly with the truly on line applications of X-ray fluorescence.

The use of conventional X-ray spectrometry for on line process control has long been recognized [22], and at the present time there are probably something like two hundred units in regular use in mineral benificiation plants in various parts of the world. Conventional X-ray spectrometry similar to that already described in previous chapters has been used for on stream applications in which the normal fixed sample is replaced by a slurry cell. Special purpose isotope excited systems have also been used with a great deal of success.

Probably the major use of on stream X-ray fluorescence analysis is in mineral extraction, and applications fall roughly into three categories:

a. Analysis of flowing slurries, found in most ore dressing and mineral beneficiation plants.
b. Analysis of dry flowing solids, for example, crushed ores on conveyor belts or dry raw cement mix en route to the kiln.
c. Analysis of drill cores and insides of bore holes.

In each of the three areas, but especially in the first, the major disadvantage of the on stream analyzer system based on a conventional X-ray source and crytal dispersion arrangement is that its high inherent cost limits the size of the installation to one or at most two units. This in turn requires some extremely sophisticated plumbing arrangements for the sample streams, leading to delays in getting the analysis to the control center. A cycling system is generally employed to allow sampling of multiple lines, but even allowing for certain priority channels, the cycle time can be of the order of 3–15 minutes. Although this may be satisfactory in some installations, the growing trend towards use of on line computer control requires almost immediate use of analytical data, so that delays of the order of minutes are unacceptable. In addition there are special problems in the determination of the lower atomic number elements ($Z < 20$), partially due to the attenuation of the fluorescent signal by the plastic windows of the flow cell, but even more due to particle effects which arise because individual particles of solid in the slurry may be heterogeneous over the same order of magnitude as the penetration depth of the X-ray beam.

The potential use of radioisotope sources for on stream analysis is also well established; a survey by Watt [23] includes an impressive list of applications and feasibility studies involving the use of radioisotopes for the on stream analysis of slurries. The problems successfully handled thus far include the determination of elements including calcium, copper, zinc, niobium, molybdenum, barium and lead in flotation feeds and slurries. A great advantage of the radioisotope system is its low cost and the fact that sources can be obtained or specially fabricated for almost any special application. Unfortunately the photon yield is generaly much lower than with X-ray tube excited sources, which precludes the use of the relatively inefficient crystal dispersive spectrometers. Energy dispersive spectrometers have played an important role in this area, and additional techniques, including use of filters and secondary sources, have done much to enhance their applicability. Typical of these special sources is the γ–X source [24]. The γ–X source comprises three essential parts: the γ-source itself (the isotope 153 Gd is frequently used), the target material in the shape of a cone, and a lead shield. The γ-source excites K X-radiation from the target cone, and this secondary X-radiation is then directed onto the sample under analysis.

Looking now at the second area, the conventional X-ray spectrometer, modified for the analysis of a flowing sample, has found some success in the analysis of dry flowing solids. As an example, use of continuous sampling on sticky tape, followed by monitoring with an energy dispersive spectrometer, has yielded some success, and detection limits of a few mg/cm^2 have been reported in the determination of heavy metals in fly ash from waste incineration [25]. For many materials, however, a major problem remains the particle size effect referred to previously. A good deal of success has nevertheless been achieved by use of successive sampling rather than continuous irradiation. In continuous irradiation a sample depth equivalent to tens of microns is continuously irradiated, for an integration time related to the number of counts required, this time being typically of the order of 60 seconds. Following a short delay for the data readout, the process is then repeated. In successive sampling, a sample is taken at time intervals whose length is equal to the time required to prepare and analyze the sample, again typically of the order of 200 seconds. In some instances the sample may be taken continuously and an aliquot taken for analysis, but in all cases, a homogeneous specimen of about 2–5 grams is analyzed. In the first case one relies on homogeneity over the cross section of the stream, but this is not critical in the second method, where a larger analyzed volume is taken and homogenized before analysis. The successive sampling technique, combined with high pressure pelletizing or borax fusion of the sample, followed by conventional X-ray fluorescence spectrometry has been applied with great success, for example, to the analysis of cements.

The third area concerns the analysis of drill cores and the insides of drill holes. Data have been reported on the direct analysis of drill core samples using PIXIE [26] as well as more conventional energy dispersive X-ray fluorescence [27]. Special instruments based on the use of semiconductor counters have been used for certain applications in the analysis of the insides of drill core holes, specifically for the determination of high atomic number elements in low grade ores. For example, Burkhalter [28] has described a system for the determination of gold and silver in a silicon matrix at concentration levels down to 20 ppm. In this example an annular source of ^{125}I was used in combination with a silicon semiconductor counter for the determination of silver and a germanium detector for the analysis of the higher energy gold K spectrum. In another example, a precision of 0.1% has been obtained in the determination of iron in dolomite bore hole samples using radioisotope X-ray fluorescence based on an energy dispersive technique using ^{109}Cd as a source [29]. This precision is sufficient as specified for oil and gas exploration.

REFERENCES

1. Slavin, W. (1986). *Anal. Chem.* **58**, 589A.
2. *Anal. Chem. Fundam. Rev.*, April 1986.
3. Hayasaka, T., Shibata, Y., Inoue, Y., Hayashi, H., and Kurosawa, Y., (1985). *Kawasaki-shi Kogai Kenkyusho Nenpo* **12**, 5.
4. Coetzee, P. P., Hoffmann, P., Speer, R., and Liesser, K. H. (1986). *Fresnius Z. Anal. Chem.* **323**, 254.
5. T. G. Dzubay, (Ed.) (1977). *X-Ray Fluorescence Analysis of Environmental Samples*, Ann Arbor Science: Ann Arbor, Michigan.

6. Kocman, V., Foley, L., and Woodger, S. C. (1985). *Adv. X-Ray Anal.* **28**, 195.

7. Cox, G. A., and Gillies, K. J. S. (1986). *Archeometry* **28**, 57.

8. Crown, P. L., Schwalbe, L. A., and London, J. R. (1985). *Adv. X-Ray Anal.* **28**, 169.

9. Stiegelschmitt, A., and Tormandl, G. (1985). *Sprechsaal* **118**, 974.

10. Yap, C. T. (1986). *Phys. Bull.* **37**, 214.

11. Yap, C. T., and Tang, S. M. (1985). *Appl. Spectrosc.* **39**, 1040.

12. Akopov, G. A., Berdikov, V. V., Zaitsev, E. A., Iokhin, B. S., and Krinitsyn, A. P. (1986). *At. Energ.* **60**, 64.

13. Frank, A. S., Schauble, M. K., and Preiss, I. L. (1986). *J. Amer. Pathol.* **122**, 421.

14. Preiss, I. L., Ptak, T., and Frank, A. S. (1986). *Nucl. Instrum. Methods Phys. Res. Sect. A* **A242**(3), 539.

15. *Analytical Chemistry, Fundamental Reviews*, American Chemical Society (published in even years).

16. *CA SELECTS—X-Ray Analysis & Spectroscopy*, Chemical Abstracts Service: Columbus, Ohio (bi-monthly).

17. *International Journal of X-Ray Spectrometry*, Wiley/Heyden: London (quarterly, 1971–).

18. Adler, I. (1966). *X-Ray Emission Spectrography in Geology*, Elsevier: Amsterdam.

19. Cesareo, R. (1982). *X-Ray Fluorescence (XRF and PIXIE) in Medicine*, Field Education Italia.

20. Herglotz, H. K., and Birks, L. S. (1978). *X-Ray Spectrometry*, Dekker: New York.

21. Callis, J. B., Illman, D. L., and Kowalski, B. R. (1987). *Anal. Chem.* **59**, 624A.

22. Jenkins, R., and de Vries, J. L. (1971). *Can. Spectr.* **16**, 3.

23. Watt, J. S. (1970). *Australian I.M.&M. Proc.* #233, 69.

24. Hope, J. A., and Watt, J. S. (1965). *Int. J. Appl. Rad. Isotopes* **16**, 9.

25. Lorber, K. E. (1986). *Waste Manage. Res.* **4**, 3.

26. Carlsson, L. E., and Akelsson, K. R. (1981). *Adv. X-Ray Anal.* **24**, 313.

27. Arthur, R. J., Laul, J. C., and Hubbard, N. (1984). *Adv. X-Ray Anal.* **28**, 189.

28. Burkhalter, P. G. (1969). *Instrum. in Industry and Geophys.* **20**, 353.

29. Tham, F. S., and Preiss, I. L. (1986). *J. Radioanal. Nucl. Chem.* **99**, 133.

INDEX

171